Poultry Farming and Keeping

Poultry Farming and Keeping

Divyesh Pandey

RANDOM PUBLICATIONS
NEW DELHI (INDIA)

Poultry Farming and Keeping

ISBN 978-93-51113-73-6

Published in 2014 in India by

RANDOM PUBLICATIONS

4376-A/4B, Gali Murari Lal, Ansari Road
New Delhi-110 002
Phone : +9111-43580356, 011-23289044, 011-43142548
e-mail: sales@randompublications.com,
info@randompublications.com, randomexports@gmail.com

Reprinted 2024

Type Setting by: Friends Media, Delhi-110089

Digitally Printed at : Replika Press Pvt. Ltd.

Preface

Poultry farming is the raising of domesticated birds such as chickens, turkeys, ducks, and geese, for the purpose of farming meat or eggs for food. Poultry are farmed in great numbers with chickens being the most numerous. More than 50 billion chickens are raised annually as a source of food, for both their meat and their eggs. Chickens raised for eggs are usually called layers while chickens raised for meat are often called broilers. In total, the UK alone consumes over 29 million eggs per day· In the US, the national organization overseeing poultry production is the Food and Drug Administration (FDA). In the UK, the national organisation is the Department for Environment, Food and Rural Affairs (Defra). According to the Worldwatch Institute, 74 percent of the world's poultry meat, and 68 percent of eggs are produced in ways that are described as 'intensive'. One alternative to intensive poultry farming is free-range farming using lower stocking densities. Friction between supporters of these two main methods of poultry farming has led to long-term issues of ethical consumerism. Opponents of intensive farming argue that it harms the environment and creates health risks, as well as abusing the animals. Advocates of intensive farming say that their highly efficient systems save land and food resources due to increased productivity, stating that the animals are looked after in state-of-the-art environmentally controlled facilities. The most intensive poultry farming methods are very efficient and allow meat and eggs to be available to the consumer in all seasons at a lower cost than free-range production· Poultry producers routinely use nationally approved medications, such as antibiotics, in feed or drinking water, to treat disease or to prevent disease outbreaks· Some FDA-approved medications are also approved for improved feed utilization.

Free-range poultry farming allows the birds to roam freely for a period of the day, although they are usually confined in sheds at night to protect them from predators or kept indoors if the weather is particularly bad. In the UK, the Department for Environment, Food and Rural Affairs (Defra) states that a free-range chicken must have daytime access to open-air runs during at least half of its life. Unlike in the United States, this definition also applies to egg laying hens. The European Union regulates marketing standards for egg farming which specifies a minimum condition for free-range eggs that "hens have continuous daytime access to open-air runs, except in the case of temporary restrictions imposed by veterinary authorities". The RSPCA" Welfare standards for laying hens and pullets" indicates that the stocking rate must not exceed 1,000 birds per hectare (10 m^2 per hen) of range available and a minimum area of overhead shade/ shelter of 8 m^2 per 1,000 hens must be provided. Free-range farming of egg-laying hens is increasing its share of the market. Defra figures indicate that 45% of eggs produced in the UK throughout 2010 were free-range, 5% were produced in barn systems and 50% from cages. This compares with 41% being free-range in 2009. Finding suitable land with adequate drainage to minimise worms and coccidial oocysts, suitable protection from prevailing winds, good ventilation, access and protection from predators can be difficult. Excess heat, cold or damp can have a harmful effect on the animals and their productivity. Unlike cage and barn systems, free-range farmers have little control over the food their animals come across, which can lead to unreliable productivity, though supplementary feeding elliminates this uncertainty. In some farms, the manure from free-range poultry can be used to benefit crops.

The texts are arranged in a lucid form and written in colloquial English. All the essential aspects of this subject have been included. Hopefully, the present study will prove very useful for students and teachers.

I thank all members of my team who have helped in the preparation of the book. My special thanks go to"Random Publications" who have published the book.

—Divyesh Pandey

Contents

1

Essential Elements in Poultry Feeding

Digestion and Metabolism

Digestion is the breaking down of the feedstuff into simple nutrients in the digestive tract in order to prepare these for absorption. Metabolism means all the changes which nutrients undergo from the lime they are absorbed into the body until they appear as excretory products. The knowledge about the manner in which the feedstuff are eaten, digested and utilised is essential for the proper understanding of practical nutrition.

Digestive Tract

The digestive tract of the fowl consists of a hollow tube running from mouth to cloaca, modifying along its length into characteristic organs namely pharynx, oesophagus, proventriculus. Gizzard, duodenum, small intestine, large intestine and cloaca. Opening into the tube and attached to it are four other organs: the crop liver, pancreas and the caeca.

Mouth/Pharynx

The mouth has a horny beak which is prehensile (capable of grasping). The tongue aids in positioning and moves the food towards the throat for swallowing. The birds have a reasonably good sense of taste with the taste buds being located in several areas of the mouth and beneath the tongue.

Oesophagus

It is a distendable tube and its role is to pass the food meterial from the mouth to the stomach. The crop serves mainly as a storage organ for food.

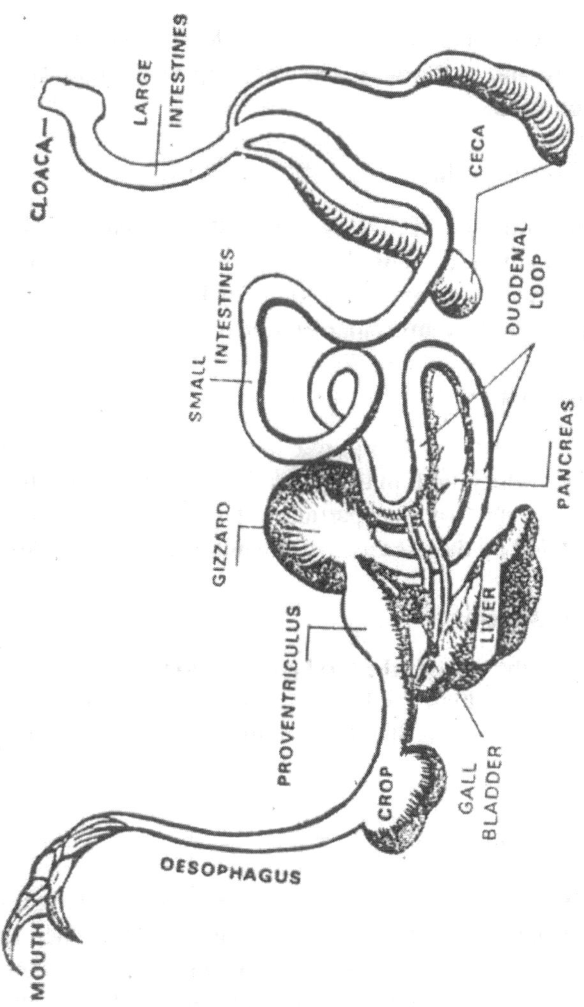

Proventriculus

Proventriculus is richly supplied with gastric glands and is sometimes referred to as the true stomach. Its storage capacity is small and docs not retain the food for long, instead leads the food into the gizzard or muscular stomach.

Gizzard

The gizzard is composed of two thick muscles opposite to each other, which help in the break down of the coarse feed or whole grain to the state of fine particles. The process of grinding whole grain is aided by the presence of insoluble grit or particles of gravels.

Duodenum

From the gizzard the food passes into the V-shaped loop known ! as duodenum which encloses pancreas. Pancreatic and bile ducts open into the gut at the junction of the deodenum and the small intestine.

Small intestine

The small intestine is five to six times the length of the bird and is structurally adapted for absorption. The inner layer of the intestine is lined with small finger like projections called villi. Each villus has a lymph capillary and a close network of food capillaries. Most of the chemical digestion as well as absorption, or the transfer i of nutrients from the gut into the blood and lymph systems, takes place here.

Large Intestine

Large intestine is an enlargement of the small intestine which extends from the point of caecal openings to the cloaca or rectum. It is relatively short and straight compared to small intestine. There is not much digestive action in the large intestine, but it acts as a storage site for food residues and a place for water resorption into the blood.

Caeca

Caeca are the two blind-ended tubes. They originate from the meeting point of the small and large intestines. They have a role in the digestion of cellulose and they are particularly well developed in goose which is an excellent forager. There is also some synthesis of vitamin B-complex and perhaps vitamin K in the caeca. The role of caeca in the domestic fowl is limited and this species utilises cellulose very poorly.

Cloaca

Cloaca is the common receptacle of the genital, digestive and urinary system and serves the triple purpose of excreting faeces, urine and eggs or seminal fluid.

Pancreas

The three lobes of the pancreas lie between the curve of the duodenum. Secretary ducts from the pancreas open into the distal end of the duodenal loop adjacent to the opening of the bile duct.

Gall Bladder and Bile Ducts

The bile duct is present on the left lobe of the liver. They open into the duodenum along with the pancreatic ducts. Bile is pushed into the duodenum due to the vigorous contractions of the gallbladder.

Process or Digestion

Fowls have a high metabolic rate. Thus it necessitates a more or less continuous supply of food to the digestive tract During dark hours when the birds are not eating, stored food in the crop is used for this purpose.

The rate of passage of feed through the alimentary canal depends to a large extent, on the feed intake which is determined by the environmental temperature, physiological state of birds, water intake and the physical and chemical nature of the diet. A single meal especially in the mash form passes through the canal faster than the ungrounded hard grains.

The breaking down process of feed nutrients (i.e. digestion) includes the following three processes:

1. Mechanical. Mechanical activities are the grinding of the larger particles of food by the action of gizzard and the muscular contraction of the alimentary canal.
2. Chemica. The main chemical action is brought about the enzymes secreted by the bird in various digestive juices. These secretions include:
    a. Saliva -secreted by the salivary glands in the mouth.
    b. Mucus -secreted by the mucus glands of the crop, which contains no enzymes.
    c. Gastric juice -secreted by the proventriculus; hydrochloric acid (HCI) contains the enzyme pepsin.
    d. Pancreatic juice -secreted by the pancreas. It is slightly acidic in nature and contains amylase, lipase and trypsin.
    e. Bile -secreted by the liver. It contains no enzyme, but it aids in emulsifying fats and thus facilitates the action of lipase.
    f. f. Intestinal juice -secreted by the small intestine. It is a source of amylase, trypsin, maltase and sucrase.
3. Microbial. Microbial digestion of food is also an enzymatic digestion but these anzymes are not secreted by the animals. These enzymes are secreted by the bacteria and microorganisms. The caeca in the birds harbour a wide range of microbial population which may aid in the digestion of the crude fibre of the diet, but as stated earlier, caeca playa limited role in the domestic fowl. However, in the case of geese, they are well developed to the level that geese can digest scales too.

In brief, the whole process of digestion consists of softening the food in the crop, mixing with digestive enzymes in the glandular stomach, grinding in the gizzard, breaking into simpler units and absorption in the intestine and finally excretion of undigested food.

Digestion, Absorption and Assimilation

The feed is picked up by the beak and swallowed without mastication. In the mouth, the feed gets mixed up with saliva which has enzymes that act on the carbohydrates present in the food. Thus 3 to 5% of all the starch that is eaten, will be broken down to maltose by the time food is swallowed. The food then reaches the crop by peristaltic action, where it gets mixed with the secretions from the crop wall and then descends to the gizzard in small quantities at a time. Meanwhile, the proteins of the feed are acted upon by ~ various gastric juices and enzymes are broken down into simpler proteins and finally to amino-acids as it reaches the small intestine. In the gizzard, the feed is well ground and mixed with digestive juices secreted by the glandular stomach. Insoluble grit, picked along with the grains are held in the gizzard and aids in the grinding process, especially when whole grain and fibrous grasses are eaten. The feed then enters the duodenum where further digestion takes place due to the action of bile from the gall bladder and enzymes from the pancreas. Here the fat present in the feed is first emulsified or broken down into smaller droplets. Most of the feed is thus acted upon and reduced to simpler substances by the time they reach the small intestine.

The digested nutrients get absorbed through the small intestinal wall and pass on to the blood stream. The blood carrying the digested material passes through the liver on its way to the heart which pumps it to the various parts of the body. The liver removes at least a part of the toxic substances that may have found their way into the blood stream. It also transforms some of the digested carbohydrates into glycogen for energy purpose in the muscles of the body. Some fat is also stored in the liver. A part of the digested feed is built into new tissues for growth and repair, some get converted into eggs and others remain stored as glycogen and fat, to meet the heat and energy requirement.

Nutrients in Feed

The feed contains substances known as nutrients, and for obtaining these nutrients to perform various functions in the body, the food is consumed daily. This has been discussed in a separate booklet No. 25 on Animal Nutrition. There are more than 40 such nutrients required by the poultry. These are arranged into six groups namely (a) water, (b) carbohydrates, (c) fats, (d) proteins, (e) vitamins and (f) minerals. The carbohydrates and fats can be further combined into the group "energy feed", as their main function is to supply energy for the maintenance, growth and production of the poultry.

Energy (Carbohydrates and Rats)

Energy is not a nutrient by itself, but a property that a nutrient possesses. Carbohydrates and fats are the main nutrient sources of energy, required by the birds. The dietary energy consumed by a bird, can be used to supply energy for various body function or be converted into fat stored in the body or given out in some form of a product. Energy consumed in excess if that which is required for normal growth and activity of the bird is stored mainly as body fat. Excess energy once ingested cannot be excreted from the body. For most efficient nutrition, the diet should contain energy in definite proportion to other nutrients needed to produce the desired growth, production of eggs or deposition of carcass fat.

The fowl obtains the required energy from the ingested feed. But, whole of the energy present in the feed is not available to them. It has been estimated that the net energy value of the diet of the chicken ranges between 70% to 90% of the total energy.

Carbohydrates can be completely replaced in the diet of chicken by fats, but the growth rate declines significantly. However, fat utilisation is very efficient in chicken and growing chicken may be fed with diets having 12% fat without any effect on the digestibility of the food. Of all the fatty acids, only linoleic acid is an essential nutrient in the diet of a chicken. Others are important mainly as the concentrated source of energy, as materials which decrease dustiness of the feed, lubricate the passage of diets and increase the palatability of certain feeds.

Of the total energy requirement, the energy required for maintenance makes a large proportion. For the body, meeting the energy requirement for maintenance takes precedence over the energy requirement for growth or egg production. For a newly hatched chick, weighing approximately 40 g, energy requirement for maintenance is about 8 kcal metabolizable energy per day. The energy required for maintenance has to be met before a young chick can utilise any dietary energy for growth. The feed consumption in the growing birds is inversely related to the dietary energy level. Birds eat more of a low energy diet than of a high energy diet, and in this way attempt to restrict their consumption at normal level. But in general, rations with higher energy content will permit fat birds, while diets having lower energy levels will produce lean chicken. During the initial phase of growth, there is little deposition of fat in the chicken, body. The carcass of broilers contains only 4% of fat during the starter phase. Chicken of the high laying, light weight strain, kept under moderate environmental temperature require 2700 kcal/kg to 3100 kcal/kg of energy levels in the ration corresponding to the moderately not conditions to moderately cool conditions.

When energy requirements for maintenance are not met through dietary sources, body energy stores are consumed up in the following order:

a. the small amount of glycogen normally stored in the liver and muscle tissues is used up;

b. body stores of fat are depleted;

c. ultimately the protein tissues are also exhausted to maintain the glucose level of blood and to permit other vital body processes.

Protein

Proteins are the major constituents of the soft tissue in the animal body. For growth, egg production, and repair of wear and tear of tissues, a continuous and adequate supply of protein in the diet of the chicken is essential. All the proteins are made up of amino-acids. However, the nature and number of amino-acids comprising various animal and plant proteins are different Plant proteins vary from each other and also from animal protein. Chicken cannot synthesize many of the amino-acids present in their body proteins. These amino-acids must be provided through dietary proteins. They are: arginine, lysine, histidine, leucine, isoleucine, valine, methionine, threonine, tryptophan and phenylalanine. Glycine is essential in the diet of the growing chicken but not in the diet of the adult birds.

High quality dietary protein should supply all the required essential amino-acids in the same proportion in which they occur in the body protein. Also the dietary protein should have a high digestibility. Diets containing such high quality proteins can fulfill the protein needs of the body with a minimum intake. In case, the dietary proteins are deficient in one or more of the essential amino- acids, they will not be able to provide proper protein nutrition even with excessive intake. The amino-acids which are not used for the synthesis of protein by the body do not go waste as food, since they are converted to an energy source. Chicken body contains various kinds of proteins which differ in their amino-acid composition.

Therefore, the usefulness of the dietary protein depends, to some extent, on the purpose for which it is fed. Fewer amino-acids are required for maintenance of the body than for the growth. Hence, a protein may be adequate in amino-acid content for meeting the maintenance needs of adults, yet it may fail to permit growth.

Certain proteins of plant origin are harmful to chicken. For example, the proteins of raw soyabean and various pulses contain certain deleterious substances which cause growth depression in the young chicken. However, these proteins are rendered harmless when soybean meal or pulses are given after proper heat treatment.

The feed consumption of the layer is influenced by the following factors:

a. energy content of the diet (ME);
b. temperature in the layer house;
c. rate of egg production;
d. feeding space per hen and infection of infestation with parasites; and
e. body weight and strain of layer.

A constant calorie : protein ratio (C:P ratio) of 173-180: 1 needs to be maintained in the diet of layers in order to ensure that the daily protein intake of birds does not suffer due to imbalance in the diet.

Diets having wider C:P ratios permit higher fat deposition on the carcass and improve its finish and can be advantageously applied to the finisher broilers. When lean meat is desired, the C:P ratio can be reduced.

The recommended C:P ratio for the diet of various classes of chicken are given below.

Starter chicken (0-8 weeks)	135:1
Grower chicken (8-20 weeks)	140:1
Layer (20 weeks onward)170-	180:1
Starter broiler chicken (0-6 weeks)	135:1
Finisher broiler (6 weeks onward)	155:1

Vitamins

Vitamins are organic substances required by the body in minute amounts for normal growth, health and production. Chicken depend on food sources for their requirement of vitamins. They are also more susceptible to vitamin deficiencies because :

a. the microorganisms in caeca of these birds do not provide any vitamin synthesis, instead compete with the host for dietary vitamins;

b. due to higher metabolic rate, chicken have a higher need for vitamins which act as the "spark plugs" of the vital metabolic reactions in the body; and

c. intensively kept chicken undergo many stresses which increase their vitamin requirements.

The following vitamins should necessarily be included in the diet of all birds. Vitamin A , Vitamin D3 (a type of vitamin D), Thiamine Riboflavin, Pentothenic acid, Nicotinic acid, Biotin, Vitamin B12, Alpha-tocopherol, Choline Vitamin C (ascorbic acid) is not considered as essential in the diet of the chicken because, under normal conditions the body synthesis of this vitamin is adequate to meet their requirements. However, under certain stress conditions, chicken are not able to synthesize optimum amounts and in such circumstances ascorbic acid supplemented diets prove beneficial. Addition of vitamin C in the diets is also reported to increase the hatchability of eggs during summer months.

Minerals

In addition to the elements like carbon, hydrogen, nitrogen, oxygen and Sulphur which comprises the organic chemical compounds of the body, chicken need at least thirteen minerals for their proper nourishment and production. These minerals are : calcium, phosphorus, sodium, potassium, chlorine, magnesium, manganese, zinc, iron, copper, molybdenum, selenium and iodine. There is increasing evidence that two more elements viz. flourine and chromium also perform essential functions in the body.

Calcium and phosphorus are found in the body mostly in combination with each other and constitute over 70% of the body ash. Deficiency of either in the diet limits the nutritive value of the other one. Growing chicken utilise most of their dietary calcium for the formation of the skeletal system. In the layers, it is utilised for the formation of egg shells, Similarly, phosphorus is important in the metabolism of carbohydrates and fats. Both calcium and phosphate , salts play an important part in the maintenance of acid -base balance. For optimum results, calcium and phosphorus contents in the diet of growing chicken should be in the ratio of 1-2.2 : 1. Dietary calcium and phosphorus in the ratio of 2.5: 1 may be tolerated by the chicken but a ratio of 3.3: 1 is decidedly injurious. It produces leg abnormalities and rickets. In practical feeding, sodium, potassium and chlorine are not I of much importance because their need is already met in their diets. However, these minerals are required regularly in the diets because even any excess consumption of it is immediately excreted and, therefore, body does not store them.

About 70% of the body magnesium is present in the skeletal system. Most of the magnesium found in the eggs is present in the shell portion. Embryo utilises 1 to 1.8 mg of the magnesium present in the cell for its development during the incubation. Chicks started on a magnesium -free diet die within a few days. Feeding of low magnesium diets to-laying hens results in rapid decrease in egg production and in growing chicks, the growth rate decreases.

Manganese content of the body is very little, but it performs many specific essential functions in the body. Common poultry feeds are not very rich in manganese therefore, it is of practical importance in their feeding. Dietary manganese is poorly absorbed in chicken. Thus, more manganese is required to be included in the chicken diets than the requirement. Rice bran, maize fibre and Lucern meal are considerable sources of manganese, but even these have to be supplemented. Manganese salts are not very expensive and poultry diets can be supplemented with manganese in an inorganic form. The other elements are required in traces and need not be \ supplemented in the diet, except in case of iodine where practical diets for chicken generally need to be supplemented with some iodine salt because the common natural feeds may not contain more than 0.3 ppm (parts per million) iodine.

Water

Animals can tolerate a loss of 98% body fat and 50% body protein but not more than 20% body water. A young growing chicken will consume water at about 20% of its body weight. Hens will consume about 14% of their body weight at 20'C and double that amount at 35'C. Also water constitutes about 85% of body weight of the chicks and 55-60% in the adult fowls. Egg constitutes 60% of its weight as water. Therefore, if the birds are denied water or given only a limited quantity it has an adverse effect on their growth rate as well as production. Prolonged deficiency of water may even lead to death. It has been experimented that if 11 day old and 18 day old chicks are not given water for 24 hours and 48 hours respectively and then suddenly supplied with sufficient water, the chicks tend to die. This is called water intoxication as it has an adverse effect on the metabolic functioning of the birds.

Water is an essential nutrient, therefore, supply of good quality water is indispensable for better growth and reproductive performance in poultry. The portable water contains many substances in solution and in suspension, some of which are mainly responsible for causing a number of problems in poultry. It may increase the morbidity, i mortality, service disease and disease conditions like Coccidiosis, Salmonellosis, Enteritis, Ascitis, Nephrosis, Septicaemia and Hepatitis. These conditions may ultimately lead to death. Thus, there is a need for complete examination of water before it is supplied to the birds. There are several methods of making water suitable for poultry drinking including chlorination, dechlorination, retention and filtration.

Feeding Various Classes of Chicken

The nutrient requirements and feeding practices of chicken depend on the purpose for which the flock is kept. The purpose of a poultry enterprises could be either for egg or meat production. For satisfactory results in this business, care has to be given to the growing chicken starting from the first day of their hatching.

Regarding feeding management, now-a-days, the diets to poultry are fed in the form of mash, pellets or crumbles. Feeding mash is common. However, pellet feeding is gaining popularity. The size of pellets for poultry ranges from 5/32" or 3/]6". With pellets, the time taken on feeding is less and the intake of feed with pellets or crumbles contributes to increased feed efficiency.

Layer Nutrition

On the basis of the growth rate and the accompanying nutrient requirements for production and growth, the layer nutrition can be divided into three phases:

1. 0 -8 weeks -chick phase
2. 8 -2Q weeks -grower phase
3. 20 weeks onward -layer.

The first period shows the most rapid growth and therefore, a high protein level (22%) is needed in the diets. The second period, the growth is less rapid and requires only 16% protein in the diets. The layers require more protein for egg production, therefore, their diet should essentially consist of 18% protein. Each of these periods are dealt with separately here.

Broiler Nutrition

Broiler is a young chicken, usually 6-12 weeks of age of either I sex specially raised for meat production. Because of the need to obtain rapid growth in broiler, their nutrient requirements are higher than the chicken being raised for egg production. On the basis of growth rate, the broiler nutrition can be divided into 2 phases:

0 -5 weeks -starter phase

6 -9 weeks -finsher stage

In practice, diets for starter broilers (0-5 weeks) should contain 3100-3400 kcal ME per kg, for maximum growth rate. Finisher diets (6th week onwards) are recommended to have energy level of 3400 kcal ME per kg feed.

Feeding of the high protein pre-starter diets is beneficial since it gives a stimulus for the early growth of the broiler. Since the quality of a protein depends upon its constituent amino-acids it is necessary to know the minimum quantities of the amino acids in the broiler diets.

C. Desi Birds Diet

Desi birds in the village home-units run on range and pickings. If there are only a few birds, they can get enough feed to pick and make up a balanced ration from waste grains, shed seeds, maggots, cow dung, bones, insect, green pickings, meat and kitchen wastes, vegetable tops and prunings, wastes at weekly markets etc. When they get good pickings, poultry pay well. But kitchen scraps and vegetable trimmings cannot meet the requirements of layers. These may be useful in supplementing other more appropriate feeds. It is essential for the profitable production of good qualily eggs that layers are fed with a diet containing the correct balance of energy, proteins, vitamins and mineral ingredients.

The desi birds should be given a small quantity of mixed ration daily to provide very cheap and balanced feeding. The village birds will combine this with feeds available in the fields when they are let out, to make a balanced feed. This way the bird will lay marc number of eggs and compared to the expenditure on feed, the profits will be more. Alternately, the birds can be given broken maze, paddy or ragi grains about 1 kg for 20-40 birds daily and permit them to run outside freely on range.

This may cost very little, but one can get , 25-50% more number of eggs that the birds on completely free range but again not as much as the same birds would give on deep litter feeding on full mash ration.

Therefore, for the backyard units of desi birds, it is better to give them some feed mixtures and then they arc let loose on the fields for supplementing the given feed. These birds have an innate capacity to pick and eat only those ingredients as are required by their nutrition status. For example, if high quality protein is missing from the diet, then the birds will selectively pick and feed on the insects or worms to balance their diet.

Given below are the examples of feed mixtures for the desi birds.

1. For chicks and growers:

Maize	50 parts
Groundnut cake	15 parts
Broken wheat	5 parts
Fine rice bran	10 parts
Fish meal	10 parts
Molasses	5 parts
Mineral mixture	
+ vitamins + antibiotic 5 parts	

2. For layers:

Maizc	60 parts
Groundnut cake	10 parts
Wheat bran	15 parts
Fish meal	10 parts
Mineral vitamins and	5 parts

Rice bran can substitute wheat bran. Pulses can substitute wheat for better results

Feeding Management

A sound feeding programe during the rearing of poultry is essential for the production of high quality chicks without much mortality.

Quantity to be Fed

Quantity of the mixed ration to be fed depends upon the age groups of poultry. For the chicks of 0.2 weeks, finely broken rice or other small grains or bread crumbs are given are the first feed. Skimmed milk or butter milk should be supplied for drinking instead of plain water all the time. After a week, chick mash is introduced which is specially prepared with finely broken grains of rice, ragi, cambu, wheat or maize, with aquatic quantities of vitamins, minerals, antibiotic feed supplements and coccidiostat.

After 2 weeks, the quantity of chick feed is increased gradually. After 6 weeks, wet mash is introduced along with succulent, well chopped greens like Lucern. Some onions and garlic well chopped may also be included.

The quantity required of chick will very from 10 g to 50 g per chick according to their age (0-8 weeks). After 8 weeks, the chick mash must be gradually substituted by grower mash, enriched with vitamins and minerals including all the essential nutrients for healthy growth at an economical cost. Quantity fed varies from 50 to 80 g per grower per day, according to age (8-20 weeks).

At about 18th week, specially prepared layer mash fully vitaminised and mineralised is introduced. Plenty of well chopped greens and water is also provided. Feeding pellets instead of mash will prove more economical and neat. 80-200 g per bird per day is fed to the bird, depending on its body weight and production capacity.

2. Time or Feeding

Regularity of feeding is also important as the birds will soon learn and expect to have their feed in the punctual time, if they are in good health. The chicks are fed 5 to 6 times a day in small quantities upto 8 weeks of age. For the backyard units, the birds should preferably be let out early in the morning (just before sunrise) after providing them with clean drinking water.

A little later, some grains about 15 g per bird can be thrown out for the birds to pick. Just before noon, some greens or vegetable pruning or kitchen trimmings can be fed, well chopped and mixed with mash. Late in the evening, some more grains similar to the early morning feed can be given. This \\'ill tempt all birds to return to their right shelter after the day's roaming, These timings can be changed to suit the individual farmers, put regularity is important in order to train the birds to adjust to these timings.

Feeding Under Stress Condition

Poultry is said to be under stress, when there is any departure from the normal routine causing inconvenience to the birds. Nutritional deficiency, diseases, extreme weather conditions, crowding, debeaking and vaccination as also the unusual noise, transporting chicks etc. are the main causes of stress in the poultry. Profits from poultry operations depend upon the elimination of as many stresses as possible.

Stress in the birds is indicated by the increase in size and functioning of a number of glands in the birds body and loss of appetite. Adult animals usually lose weight while growth slows down or ceases in growing animals.

Antibiotics and vitamin A are beneficial in reducing early chick mortality in diseases and also in times of stress. Their use is suggested after any shock or upset conditions such as overheating, chilling, vaccination, deworming and debeaking. They are fed 2 or 3 days prior to the period of stress and 2 to 5 days following stresses, depending on the condition of the birds. Nutrition under heat stress deserves a special mention. Energy requirement of chicken decreases the feed consumption. The fall in feed consumption may cause general or specific nutrient deficiency. The real temperature range for layer chickens lies between 13 to 30'C. Thus the feed intake may decrease by 1.5% per 1° C increase in temperature as the latter increases from 20'C to 30'C, while the decline in feed intake may be 4-5% for each 1°C rise in temperature in the range of 30-40 degree centigrade.

While the increase in temperature decreases the energy requirement, the requirement for protein, minerals and vitamins do not decrease. The summer diet of birds should contain higher level of these nutrients.

The heat stress can be reduced by feeding diet which will lower the heat production in the body. Among the nutrients utilised by the body for energy, fats cause the lowest heat production and the proteins the maximum. Therefore, the dietary protein requirement can be minimized by ensuring that only essential amino-acids needs are met, and dietary excess of amino-acids are avoided.

Reports indicate a reduction in the body synthesis of vitamin C by birds at higher temperature. Restoration of impaired thyroid functioning (due 10 excess heat) and improvement in egg production and shell quality of eggs have been reported with the vitamin C supplementation of summer diets.

Birds require more water at higher temperatures. Plenty of cool and clean water must be ensured during summer months.

Conclusion

Poultry farmer's success mostly depends on the type, source and quality of feed, which accounts for the 70% of the total cost of egg production and 55% of the cost of broiler production. Availability of feeds and their ingredients contributed significantly to the increased poultry production of the country during the last two-three decades. Fortunately, requirements for many nutrient., for poultry are known and standards are available as a guide. However, the biggest constraint in feeding of the poultry is that, the bulk of the cereals produced in our country has necessarily to be diverted for human consumption. As a result, the main problem will be balancing the various nutrients in an available from to give the necessary Level of dietary energy, amino-acids and other nutrients necessary for the birds. Majority of the non-conventional feed ingredients are unsuited for inclusion in the formulated feeds because of the presence of excess of some deleterious factors. Some of these are unpalatable and others are incompatible with oilier feed ingredients when mixed with them.

No individual feed ingredients can supply all the nutrients required for growth and production. Therefore, a variety of food ingredients from different sources have to be mixed in such a way that it provides a balanced nutrition in an available form. For this, one has to take into account a number of factors like the breed, environment management practices, health status, the choice of ingredients and their relative cost suitable changes have to be made sometimes to suit the economic and other conditions.

Feed Evaluation

Feed evaluation is the testing of feed quality, providing information on the composition of feed or feed ingredients as well as their suitability for poultry. Poultry feed is made up of many ingredients, which are broadly grouped into providers of energy (fats, oils and carbohydrates), protein (amino acids), vitamins, minerals and product quality enhancement. Typically, cereals such as wheat, barley, sorghum and maize will provide energy while soybeans, lupins, canola and peanuts provide protein. These ingredients are then combined in such a way as to provide the energy, protein, vitamin and mineral requirements for poultry through the process of feed formulation. In order to know what amounts of these ingredients should be included in the diet, the ingredients are first evaluated, to see what nutrients they contain in what quantities. After the diet has been prepared, it may also be necessary to evaluate the complete product, to determine its suitability for the class of poultry that will be fed (such as egg layers, meat chickens or breeders).

Feed evaluation is a key process in the poultry industry. Feed ingredients need to be tested in order to formulate the complete diet, and diets have to be evaluated to determine their suitability for poultry. Evaluation provides different types of information, as required by nutritionists and farmers. In general, the range of tests that can now be performed is wide and it is now possible to obtain results rapidly.

Measures of Feed Quality

Feeds and feed ingredients can be evaluated physically as well as chemically. The physical evaluation of feed mostly provides preliminary information on the quality of the material. It involves assessing physical qualities such as weight, colour, smell and whether the material has suffered from any contamination by other materials. Chemically, feed is made up of water and dry matter. The dry matter contains organic and inorganic compounds. The organic part of feed is made of mainly carbohydrates, proteins, vitamins and fats and oils. The inorganic part is made of mineral elements, also known as ash. Feed or feed ingredients can be analysed to provide values of each of these components. Apart from obtaining values of chemical composition, the extent of utilisation of these components by the bird, termed nutritive value, is also measured.

Feed quality is measured by chemically breaking up the food into the components mentioned above. In the industry, it is sometimes necessary to break down these large components into smaller analytical fractions. Thus, values of starch and the non-starch component (called fibre) of carbohydrates may be provided. Proteins are made of amino acids, 10 of which must be present in poultry diets, so their amounts should be indicated during feed evaluation.

In the past, feed evaluation was a cumbersome process, requiring days to complete. However, newer equipment and procedures have been developed, which enable the rapid evaluation of most materials. For example, starch is determined using a ready-to-use kit and protein is rapidly determined on Leco® machines, which eliminate time-consuming digestion of feed with strong acids and reaction of material with acids and bases. Near-infrared reflectance spectroscopy is one of the latest techniques by which feed can be evaluated with the most minimal preparation of the sample.

Of greater importance in feed evaluation is the response of poultry to particular feeds. This is regarded as the real nutritive value of the feed, and must be measured as part of feed evaluation. Nutritive value does not necessarily entail animal growth or egg production. It gives information on how much of each of the fractions in feed, i.e. starch or energy, protein (amino acids), fats, minerals or vitamins, was used by the bird. When feed is given to poultry, they are able to break down only a fraction of the feed and absorb it into the body for growth and egg production. The rest is voided in faeces and urine, which are excreted together by poultry. The amount of nutrients retained by the bird is an indication of the nutritive value of the feed.

Importance of Feed Evaluation

Feed evaluation is important because ingredients that belong to the same class contain different nutrients; for example, maize provides more energy than wheat while soybeans contain more proteins than lupins. The same ingredient varies from one supplier to the other, and between years. In drought years, cereals fill poorly and are therefore lower in quality. Most importantly, if feeds are not evaluated, it is not possible to tell if the material will be suitable for feeding poultry. Feeding standards have already been set for different types of poultry, so the requirements for different nutrients must be met precisely. It is possible, with the current state of knowledge, to predict poultry growth or egg production by modelling feed quality, type of housing, class of poultry and duration of feeding. The central key issue in these models is feed quality, which can only be obtained through feed evaluation.

Advances in feeding programs for laying hens

Progressive Advances in Nutrition and Computer Technology have Developed Feeding Programs over the Last 60 Years

Feeding programs have undergone increasing complexity and sophistication since the 1950s following progressive advances in nutrition and computer technology. Applying accumulated knowledge to the needs of specific production segments of egg production has facilitated a relative reduction in feed cost expressed per dozen eggs or pound of liquid.

Development of Feeding Programs for Laying Hens

Until the 1950s, flocks held on small-scale farms were generally fed a single diet for their entire laying cycle. Feeds were generally purchased from local mills or affiliates of national suppliers. Diets were formulated on the basis of margin, which could be generated on sales given available ingredients and their cost. This is the simplest feeding program to implement. However, protein and hence amino acid intake could not be adjusted with age or feed intake resulting in excess cost of protein and reduced shell quality due to excessive egg size.

During the 1960s phase feeding programs were introduced. Dietary specifications were based on age and production. Feeding less protein as egg mass declines helps control egg size, improves shell quality and has the potential to reduce protein cost by up to 2 cents per dozen or more. Phase feeding requires a wide range of diets and offers no means to control protein intake with changes in feed consumption.

Phase feeding based on intake followed in the 1970s in responding to the reality that intake is influenced by house temperature. By adjusting dietary protein levels as feed consumption changed, protein intake could be held constant. Feeding based on intake in addition to phase feeding based on age provides even more control over egg size and shell quality and reduces protein cost by up to 4 cents per dozen or more. Successful implementation of phase feeding requires effort, skill and a range of diets along with accurate records of performance and intake.

Prior to the 1990s crude protein, or in some case amino acid requirements, were set by feeding graded levels of nutrients and specifying the amount that gave the best performance and feed efficiency. Since the value of eggs or price of feed have no influence on nutrient requirements for performance, neither cost nor revenue were generally not considered in setting nutritional requirements and any consideration of cost efficiency was generally disfavoured by journal editors.

Econometric feeding and management (EM&F) programs were introduced in the 1990s. This innovation diverged from the classic approach to feeding for optimum performance to introduce the concept of cost per dozen eggs or pound of liquid produced. Egg and liquid prices and feed cost are integrated into programs which allocate a range of available ingredients among specific diets representing phases and intake levels. The approach allowed producers to see the effect of changing egg and feed prices on protein requirements and helped to more accurately implement phase feeding programs based on intake.

Being able to quantify how returns varied as feed and egg prices changed gave producers an additional criterion to consider when selecting diets contributing to a potential increase of up to 2 cents per dozen or more in addition to the value obtained by phase feeding based only on intake. Producers could also appreciate the spread in egg price due to size and the effect on feed price due to varying energy and protein cost ratios which were just as important in determining protein requirements for optimal returns as absolute egg and feed prices.

In 2007 advanced EF&M programs were introduced that integrated phase feeding programs based on intake, a least cost feed formulation program, an EF&M program, a record keeping program, and a production control program into a single package.

This allowed producers to evaluate:

- The protein and energy requirement for optimal performance and feed efficiency since egg and feed prices are not considered in a traditional feeding program.
- The protein and energy requirement for optimal returns considering both egg and feed prices.
- The cost associated with over-or under-feeding protein as egg and feed price change applying EF&M feeding.
- The cost associated with over or under feeding energy as egg and feed price varies when feed allocation is based on intake.
- The cost associated with over- or under-feeding protein and energy as egg and feed price changes with traditional phase feeding.

Because protein and energy represent up to 85% of feed cost and 60% or more of variable production cost, it is important to integrate and summarize the above information along with flock performance criteria needed to select diets based on phase and intake. The integration of the five programs into a single package simplifies feeding based on intake while at the same time helping to optimize performance and return. Knowing the cost associated with over- or under-feeding protein and or energy in written form as feed and egg prices change is not the sole factor in selecting diets with specific protein levels. It is only one of the many criteria including egg production and case weight, hen age and weight, feed consumption, house temperature, grade requirements of the market. EF&M is used to optimize protein efficiency, flock performance and returns when phase feeding is based on intake.

Econometric Feeding and Management Programs for Egg Products

The introduction of an EF&M program for the breaker segment of the egg industry in 2010 showed the protein (or specific amino acid) and energy levels required for optimal performance and return as feed and liquid egg prices changed. Costs associated with over- or under-consumption of protein and or energy are shown in cents per pound of liquid egg and dollar value per million hens per week. One cent per dozen. spread over a million hens represents $250,000/year. Because feeding can easily influence returns by up to 6 cents per dozen appropriate selection of diets and feed allocation are significant to return. Feeding programs and feed formulation are equally important in optimizing performance and profit and requires integrated software packages.

Feeding Broiler Breeders for Chick Quality

For successful broiler production a chick requires good bodyweight, with excellent nutritional reserves at day old. It needs to be in excellent health with a fully functioning immune system. From this starting point, providing the broiler with suitable environment and nutrition will enable optimal performance to be achieved. The developing embryo and the hatched chick are completely dependent for their growth and development on nutrients deposited in the egg. Consequently the physiological status of the chick at hatching is greatly influenced by the nutrition of the breeder hen.

In reviewing breeder nutrition, it should be remembered that nutrient supply to the broiler breeder is a sum of two parts, namely nutrient content of the diet and quantity of feed supplied to the breeder birds. Both parts need to be balanced to ensure correct daily nutrient supply. It is also very important to realise that the cost of feeding the breeder appropriately to ensure good nutritional status of the chick is very low when viewed on a per chick basis and compared with the total feed cost of raising a broiler to slaughter weight. Calini (2006) calculated that the cost of breeder feed contributing to the production of a chick is equivalent to only 7% of the total feed cost for a broiler grown to 2.5Kg. This illustrates the value of ensuring the best possible nutrition of the breeder.

Nutrient Levels in Broiler Breeder Feeds

When considering nutrient levels in breeder feeds, the nutritionist must focus on the daily supply of individual nutrients to the bird. Starting with protein, studies have shown that the protein levels fed to breeders in production can affect chick bodyweight and final broiler performance. The relationship between protein content of breeder feed and chick weight seems well defined.

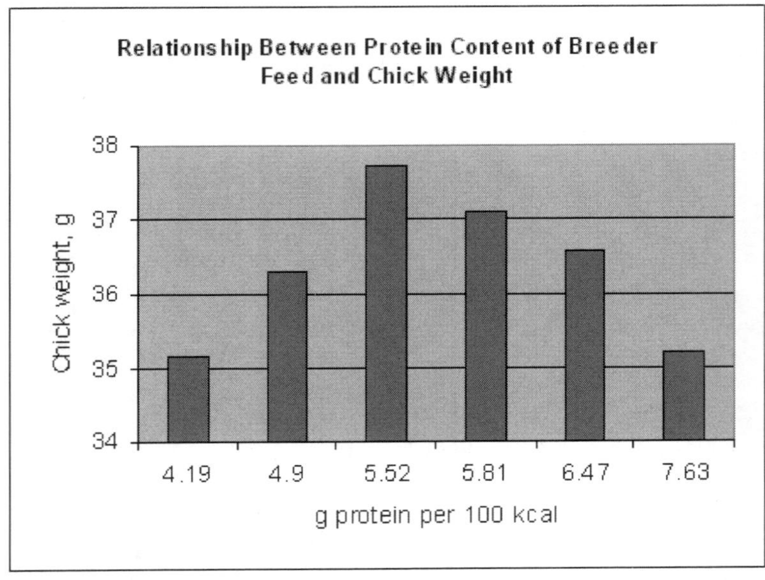

Using this information, a breeder diet with an energy density of approximately 2750 Kcal/Kg should have a protein content of 15%. This optimum protein level has been supported by other work, and it is important to remember this is an optimum level, not a minimum, as excess protein can be as detrimental as insufficient protein. In particular, it has been shown that excess protein reduces fertility. Furthermore, consideration must be given to protein quality and the nutritionist must ensure a balance of amino acids is supplied from good quality protein sources.

The impact of energy content of the breeder feed is not as well defined as that of protein. Reviewing studies carried out to evaluate optimum energy intake would suggest that 440 - 480 Kcals/bird/day is most appropriate for optimal chick quality. This equates to 160 - 175 g/bird/day at 2750 Kcal/Kg feed. When considering energy, attention must also be given to fat composition and in particular to the requirement for unsaturated fats such as linoleic acid. This essential fatty acid is required for cell membrane integrity, immune competence and embryonic development, therefore directly affecting chick quality. In practical terms, the inclusion of added fats into breeder feeds should be kept low, with preference for unsaturated fats rather than saturated fats.

The major minerals, especially calcium, phosphorous, sodium, potassium, magnesium and chloride are involved in shell formation; improvements in shell quality generally lead to better egg and chick quality. Variations in maternal phosphorous supply have been shown to influence bone ash of young but not older progeny. It is important to supply adequate phosphorus in breeder diets to ensure best possible bone integrity in the early stages of chick growth. In terms of trace minerals, most interest in this field has centred on the use of chelated minerals which have been shown to increase deposition in the egg and transfer to the tissues of the hen and the embryo. Most recent work has focused on the antioxidant status of breeders, embryos, offspring and the role of selenium. Seleno-methionine has been shown to improve both the vitamin E and antioxidative status of eggs, embryos and chicks up to 10 days of age. Supplemental zinc methionine and manganese amino acid complexes have shown improvements in chick immunity and liveability.

Table: *is a summary of those minerals which when fed to breeders have an effect on progeny performance.*

Summary of minerals fed to breeders shown to have an effect on progeny performance.				
	Growth	***Liveability***	***Immune Function***	***Skeletal***
Fluoride				X
Phosphorous				X
Selenium		X		
Selenomethionine	X		X	
Zinc	X		X	X
Zn-Methionine		X	X	

Vitamins are involved in most metabolic processes and are an integral part of foetal development, therefore the consequence of suboptimal levels of these nutrients in commercial diets are known to result in negative responses to both parent and offspring performance. Vitamins account for about 4% of the cost of a breeder feed, so economising on vitamin inclusion rates is rarely a sensible option. Generally there is a shortage of information on vitamin requirements of broiler breeders especially when related to offspring performance. Most of the breeder work is quite dated and since that time breeder performance has changed.

A review of work on fat soluble vitamins, biotin and pantothenic acid has shown that vitamin E has the largest impact on progeny. In general it seems to be justified to supplement practical breeder feeds with 100 mg/kg vitamin E. The influence of increased vitamin levels fed to young parent stock on progeny performance is an area which has received significant commercial interest. Internal and field trials have shown that increased vitamin levels (mainly B Vitamins and Vitamin E) improved liveability and early growth. A practical basis for making recommendations is to feed vitamin levels that maximise the resulting level in the egg.

The Influence of Feed Allocation on Chick Quality

Underfeeding the hen can have an impact on chick quality and this is particularly noticeable in the early production period. Modern hybrid parent flocks commence production at a faster rate than in the past and consequently egg output increases over a shorter time span during the early laying period. Feed allocations during this period have not necessarily increased in line with this egg production trend. Low feed allocation intake by young commercial breeder flocks have been shown to compromise nutrient transfer to the egg, resulting in increased late embryonic death, poorer chick viability and uniformity (Aviagen Ltd 2002). In a study by Leeson (2004) broiler breeders were fed different levels of feed through peak production varying from 140 to 175 grams. Although the increased feed allocation increased bodyweight there was no influence on egg size up to 175g, however chick weight was influenced by feed allocation.

Table: *The effects of breeder feed levels on chick weight*

Peak breeder feed (g/b/d)	*30 week breeder chick weight (g)*
140	40.3
147	40.0
155	41.5
162	41.7
169	41.8
175	42.0

Summary

Research shows that nutrient supply to the broiler breeder is of consequence to chick quality and production performance. This places greater emphasis on the nutritionist providing the correct nutrient density diet and the flock manager to provide appropriate feed intake to the bird coming into lay and through the production period.

4 Key Points

- Breeder Nutrition influences chick quality and broiler performance
- Broiler performance can be economically improved by investing in breeder nutrition
- Supply diets with adequate and consistent nutrient levels to breeders
- Manage feed quantities with reference to breeder physiological requirements, productive status and bodyweight

Broiler Breeder Nutrition and Feed Management Affects Broiler Progeny Performance

Abstract: A series of experiments were conducted to test the hypothesis that the feeding management applied to broiler parent stock (broiler breeders) could alter the performance of the broiler progeny. The broiler breeders were reared in typical litter floor black-out facilities and were moved to slat-litter breeder facilities and photostimulated at 21-22 weeks of age.

In Experiment 1 broiler breeder males were reared on two different cumulative feeding programs providing either 29,580 kcal of ME and 1,470 g of CP or 33,500 kcal of ME and 1,730 g of CP to produce either a Low body weight (Low BW) or High body weight (High BW) at 21 weeks of age, respectively. From 21 weeks of age the males were fed a common diet in the same daily quantity irrespective of the fact that there was a difference in body weight at 21 weeks of age. These males were weighed individually on a regular basis to monitor this effect. Broiler chicks were hatched from eggs collected at 29 weeks of age and grown to 42 days of age. The Low BW males exhibited numerically better fertility ($P<0.10$) and improved broiler body weight ($P<0.06$) at 29 weeks of age. As might be expected, the transition to the same feed allocation during the breeding period caused a transient plateau in body weight gain from 22 to 32 weeks of age for the High BW males. Examination of the individual body weight data revealed that it was only the largest males in the High BW treatment that experienced this failure to consistently gain body weight. Thus, the larger males with the superior genetic potential evidently did not mate and produce progeny.

In Experiment 2, two male feeding programs were applied during the production period (Constant or Increasing) beginning at 26 weeks of age. Broilers were hatched at 50 week of age and grown to 42 days of age to evaluate the effect of male treatments on progeny performance. The Constant program elicited lower broiler breeder fertility from 36 to 55 wk of age and this resulted in a lower broiler progeny BW and poorer adjusted feed conversion at 42 d of age. In this case, the larger males in the Increase treatment exhibited a consistent body weight increase while the larger males in the Constant treatment did not. Increasing male feed allocation during the production period improved fertility and favourably impacted progeny performance because the larger males continued to mate.

To evaluate the vertical effects of cumulative nutrition during broiler breeder pullet rearing on performance of broiler offspring, broiler chicks were hatched from broiler breeder females reared on a range of cumulative nutrition in three trials that used graded levels of cumulative crude protein (CP) and metabolizable energy (ME) intakes (High (27,788 kcal ME and 1,485 g CP), Medium (26,020 kcal ME and 1,391 g CP), Low (24,242 kcal ME and 1,296 g CP)) to 22 weeks of age. Breeder reproductive performance was not significantly affected. The high plane of cumulative breeder pullet nutrition increased 21-day broiler male BW in Broiler Trials 1 ($P<0.05$), 2 ($P<0.06$), and 4 ($P<0.08$) at 27, 28, and 33 weeks of age, respectively. There was no significant effect in Broiler Trial 3 when the breeders were 39 weeks of age. Greater broiler breeder pullet nutrition during rearing appeared to positively affect subsequent male broiler growth from early lay broiler breeders without significant impact on breeder performance.

Although quantitative feed restriction of broiler parent stock (broiler breeders) has long been known to be required to prevent the most obvious effect of the continued improvement in broiler genetics, excess body weight, it has not been as apparent that this obligatory management might also alter broiler progeny performance in a negative manner. It has become clear that improvements in broiler feed conversion are reducing the amount of feed required to rear a broiler breeder to a given body weight, but selection for growth rate, feed efficiency, and conformation on relatively high density diets has apparently not decreased the nutrient requirements for proper sexual development in a proportionate manner (Walsh, 1996). In fact, it is tempting to suggest that many reproductive problems have been simply due to insufficient nutrition given the genetic and nutritional background of the primary breeding stock (Siegel and Dunnington, 1985).

This was also suggested by the work of Sorneson (1980; 1985) who selected meat type chickens for increased growth rate on high and low protein diets. When tested on the low protein diet, the low protein selected line progeny grew substantially faster than the high protein line progeny. This meant that the plane of nutrition at the parent stock level would influence the required plane of nutrition at the progeny level. The implications of the protein component of this process on parent stock were further delineated by the work of Lilburn et al. (1992) who reared a heavy weight quail line, selected for several generations on a 28% CP diet, to sexual maturity on a standard 24% CP diet known to produce sexual maturity at 42 days of age in a random bred control line. Sexual maturity was delayed in the heavy weight line when fed the 24% CP diet (a diet that was suggested by the NRC to be adequate) but the delay in sexual maturity was significantly decreased by rearing the heavy weight line birds on a 30% CP diet, which was more similar to the selection diet.

It must be remembered that one of the first problems that arose in modern broiler breeders was delayed onset of sexual maturity. Subsequently, our laboratory demonstrated a specific cumulative protein need for rearing pullets, irrespective of body weight, that seriously called into question the paradigmal view of a need for absolute control of body weight relative to published reference standards provided by primary breeding companies (Walsh and Brake, 1997, 1999). Thus, it must be recognised that broiler parent stock (broiler breeders) females and males continue to exhibit similar improved conformation, growth rate, and feed conversion. This has created broiler breeders that can easily reach their standard (target) body weights during rearing even though they were grown without sufficient nutrition to reproduce and live optimally through the entire production cycle. Many long-standing paradigms need to be reconsidered.

The results of the broiler progeny have long been recognised as being as important as fertility and egg production as outcomes of good broiler breeder management but data to clearly relate broiler breeder management to broiler performance have been scarce. Less than optimum broiler performance has been observed frequently during the early and late stages of production by commercial broiler breeder flocks. Therefore, it was deemed important to investigate how the management of broiler breeder males and females could affect the performance of the broiler progeny produced by broiler breeder flocks during the early and late stages of production.

Feeding Management of Broiler Breeder Males

In order to study how the feeding management of broiler breeder males could affect broiler progeny a series of studies were undertaken. Broiler breeders were reared sex-separate in a typical black-out house to 21 weeks of age in the presence of 23 hours of light for 7 days followed by 8 hours of light daily. Birds were moved at 21 weeks of age to a curtain-sided, two-thirds slat and litter house where the photoperiod was extended with artificial light to 14 h and 15 h at 21 and 23 weeks of age, and to 15.5 and 16 h at 5% and 50% rate of lay, respectively. All broiler breeder females received a starter diet from 0 to 2 weeks followed by a grower diet to 24 weeks and a breeder diet thereafter. The separate-reared males were managed such that very similar body weight distributions were present in each replicate pen when selected and mixed with replicate pens of females at 21 weeks of age. This was to insure that the normal range of male body weights (therefore similar range of genetic potential) was fairly represented within each replicate breeding pen, and therefore, within each treatment.

In Experiment 1 broiler breeder males were reared on two different cumulative feeding programs providing either 29,580 kcal of ME and 1,470 g of CP or 33,500 kcal of ME and 1,730 g of CP to produce either a Low or High body weight (BW) at 21 weeks of age, respectively (Romero-Sanchez et al., 2007b). From 21 weeks of age the males were fed a common diet in the same daily quantity irrespective of the fact that there was a difference in body weight at 21 weeks of age. This intentionally underfed some larger males and males were weighed individually on a regular basis to monitor this effect. Broiler chicks were hatched from eggs collected at 29 weeks of age. There were 15 male chicks allocated to each pen in a 32-pen house with 16 replicate pens per breeder feeding treatment.

In Experiment 2 broiler breeder males were managed as described by Romero-Sanchez et al. (2007a). A cumulative intake of 31,460 kcal of ME and 1,669 g of CP was attained at 21 weeks of age for all males. Subsequently, two different feed allocation programs (Constant or Increase) were applied after 26 weeks of age (Romero-Sanchez et al., 2007c). The Constant feeding program maintained 110 g/male/d throughout the production period while the Increase feeding program provided biweekly increments of 1 g from 26 to 42 weeks and a similar increase every 4 weeks from 42 to 62 weeks until a daily feed intake of 123 g/male/d was reached. Broiler chicks were hatched from eggs collected at 48 weeks of age. There were 18 replicate pens of either 15 male or female chicks from each breeder feeding treatment within a 72-pen house.

The effect of male broiler breeder feeding program at 29 and 48 weeks of age on fertility and broiler progeny performance in Experiments 1 and 2, respectively, is shown in Table 1. In Experiment 1 the Low BW males exhibited numerically better fertility (P<0.10) and improved broiler body weight (P<0.06) at 29 weeks of age. There had been differences in broiler breeder male body weight during rearing but these differences diminished as the males began to be provided the same daily feed allocation, as would be expected. However, the transition to the same feed allocation during the breeding period did cause a transient plateau in body weight from 22 to 32 weeks of age for the High body weight males. Examination of the individual body weight data taken during this period of time revealed that it was only the largest males in the High body weight treatment that experienced this failure to consistently gain body weight. Fertility has been found to be largely a function of the number of males mating in a flock. In the case of the High body weight treatment the larger males were most likely not mating, as evidenced by a failure to gain body weight, and this resulted in lower broiler body weight. The logic for this explanation lies in the simple fact that the larger males in each treatment were most probably genetically similar so that when the largest males in the respective treatments did not mate the largest broilers were not produced. The larger males in the Low body weight treatment did not exhibit this effect as the daily allocation of ME was evidently sufficient to support normal growth and mating activity. This was most likely due to less than adequate ME to support their higher body weight. The smaller males in both treatments exhibited consistent body weight gain and presumably were equally efficient at mating.

Table: *Effect of male broiler breeder body weight (BW) and feeding program on fertility and broiler performance (adapted from Romero-Sanchez et al. 2008).*

Experiment	*Breeder Flock Age*	*Breeder Treatment*	*Fertility*	*Broiler*	
				BW@42 days	*Sex*
	(week)		(%)	(g)	
1	29	Low BW1	97.6x	2766x	Male
		High BW	96.2y	2669y	Male
2	48	Constant2	92.9b	2546b	Mix
		Increase	95.2a	2604a	Mix

x,y Means in a column within an experiment with no common superscripts differ at P < 0.10. a,b Means in a column within an experiment with no common superscripts differ at P < 0.05.

1 Low BW males were 3096 g and High BW males were 3,565 g at 21 weeks of age but were fed in a similar manner thereafter in Experiment 1. 2 The Constant treatment males received 110 g of feed daily from 26 to 50 weeks of age while the Increase treatment males experienced an increase from 110 to 121 during the same period in Experiment 2. The Increase feeding program during the production period produced improved broiler breeder fertility and greater broiler body weight at 48 weeks of age in Experiment 2 (Table 1). In Experiment 2 the body weight gain of the Increase treatment was greater than for the Constant treatment. Again, examination of the body weight data from individual males revealed that it was only the larger males that were the most affected by the apparently less than adequate feed (ME) allocation of the Constant versus Increase feeding programs. The failure to gain body weight in a consistent and appropriate manner was most likely the cause of the lower fertility of the Constant treatment as the larger males did not mate, as evidenced by lower broiler body weight. The data of Experiment 2 were similar to the reports of Attia et al. (1993; 1995) that showed improved broiler performance due to a greater daily ME allocation in Ross 344 broiler breeder males. {mosimage}

The conclusion appears to be simple. The broiler breeder males that produced the progeny with the greatest genetic potential were the largest broiler breeder males and they had to be fed in a programmed manner that achieved early sexual maturity and persistent fertility. Further, if fertility decreased due to less than adequate male ME allocation then there was a decrease in broiler body weight. This clearly demonstrated that the larger males in the broiler breeder flock produced the larger broilers and that they have to be fed appropriately in order to obtain maximum broiler performance.

Feeding Management of Broiler Breeder Females

To evaluate vertical effects of cumulative nutrition during the pullet rearing period on performance of broiler progeny, four broiler trials were conducted using chicks hatched from broiler breeder females reared on a range of cumulative nutrition in three consecutive broiler breeder trials. Breeder Trials 1 and 2 each used three graded levels of cumulative crude protein (CP and metabolizable energy (ME) intakes

(High (27,788 kcal ME and 1,485 g CP), Medium (26,020 kcal ME and 1,391 g CP), and Low (24,242 kcal ME and 1,296 g CP)) for pullets to 22 weeks of age. Breeder Trial 3 used only the High and Low levels. The Low feeding program could be described as concave from 2 to 22 weeks of age while the High feeding program could be described as convex in form. The Medium program was linear and fell between the Low and High programs. Males were grown sex-separate on a 17% CP, 2.90 kcal/g ME growing diet to a cumulative nutrient intake of about 32,000 kcal ME and 1,600 grams CP as determined by our laboratory to be satisfactory (Peak, 1996; 2001).

A single 16% CP, 2.90 kcal/g ME breeder laying diet and identical management practices were applied to breeder hens in all three breeder trials during the laying period. Males and females were fed separately the same breeder laying diet during the laying period. There were four replicate pens for each cumulative pullet nutrition treatment. The breeder facility was a two-thirds slat design with curtains and fans for ventilation. Breeders were moved from a black-out rearing (8 hours of light) facility to the laying facility and photostimulated (14 hours of light) at 22 weeks of age. Identity of breeder treatments was preserved during egg collection, egg storage, egg incubation, and chick processing. Broiler Trial 1 used chicks hatched from Breeder Trial 1 at 27 weeks of age. Broiler Trials 2 and 3 evaluated chicks hatched from Breeder Trial 2 at 28 and 39 weeks of age, respectively, and Broiler Trial 4 used chicks hatched from Breeder Trial 3 at 33 weeks of age. There were 12 pens of male and 12 pens of female broiler chicks from each breeder pullet cumulative nutrition treatment in Broiler Trials 1, 2, and 3 while there were 18 pens per sex in Broiler Trial 4.

Breeder Trial	*Breeder Age*	*Cumulative Pullet Nutrition1*			
		High	*Medium*	*Low*	
	(weeks)	----------------	(g)	----------------	
1	28	57.2	56.4	56.4	
1	34	63.9	62.9	63.9	
1	46	68.3	68.2	68.4	
2	28	55.3	57.2	54.5	
2	34	63.8	63.2	62.7	
2	40	67.4	67.0	66.7	
3	28	58.6	-	57.4	
3	34	64.8a	-	63.7b	
3	40	66.6		66.2	

a,b Means in a column within a trial with no common superscripts differ at $P < 0.05$.

1Breeder Trials 1 and 2 each used three graded levels of cumulative crude protein (CP) and metabolizable energy (ME) intakes (High (27,788 kcal ME and 1,485 g CP), Medium (26,020 kcal ME and 1,391 g CP), Low (24,242 kcal ME and 1,296 g CP)) to 22 weeks of age. Breeder Trial 3 used only the High and Low levels.

There was no consistent effect of cumulative pullet nutrition on any reproductive variable measured. Egg production, fertility, and fertile hatchability were very similar and quite acceptable while percentage shell was not significantly affected.

Egg weight was not significantly affected except at 34 weeks of age in Breeder Trial 3 where the High treatment exhibited greater egg weight.

All pullets were weighed at 22 weeks of age to confirm treatment effects and the Low, Medium, and High treatments typically weighed 2,450 grams, 2,550 grams, and 2,660 grams, respectively.

Although pullet body weight did differ during rearing, these differences gradually disappeared during the laying period, as would be expected once a similar daily allocation of ME was applied. Body weights taken from 20 hens per pen near the time of each egg collection for each Broiler Trial are shown in Table below.

Table: *Effect of cumulative pullet nutrition during the rearing period on subsequent breeder hen body weight1*

Broiler Trial	*Breeder Trial*	*Breeder Age*	*Cumulative Pullet Nutrition*		
			High	Medium	Low
		(weeks)	---------------- (kg) -------------		
1	1	26	3.2	3.3	3.3
2	2	28	3.6	3.6	3.5
3	2	40	4.0	4.0	3.9
4	3	32	3.8	-	3.8

1Hen body weights taken from a random sample of 20 hens from each of twelve 200-hen pens at the ages shown.

There were no consistent significant effects of plane of pullet cumulative nutrition on broiler feed conversion or mortality to 21 days of age (data not shown). However, there were reasonably consistent effects on broiler body weight as shown in Table 4.

The high plane of cumulative breeder nutrition increased 21 day broiler male BW in Broiler Trials 1, 2, and 4. There was no effect in Broiler Trial 3 when the breeders were 39 weeks of age, as might be expected of a so-called "prime" flock, although there was a residual numerical effect. Female BW was affected by plane of breeder nutrition in Broiler Trial 2 only as the effect appeared to be less consistent for female offspring.

It was interesting to note that the male body weights at 21 days of age were in the 900 gram range. This was very good early broiler growth by most standards and the fact that the increased body weight observed in the male was evidenced in the presence of such good control (Low) performance was remarkable.

Table: *Effect of cumulative pullet nutrition during the rearing period on subsequent 21-day broiler body weights*

Broiler Trial	*Breeder Trial*	*Breeder Age*	*Broiler Sex*	*Cumulative Pullet Nutrition1*			*P =*
				High	*Medium*	*Low*	
		(weeks)		-------------	(g)	--------------	
1	1	27	M	910	850	900	0.01
1	1	27	F	810	810	800	0.93
2	2	28	M	950	940	900	0.06
2	2	28	F	900	870	860	0.03
3	2	39	M	950	920	920	0.31
3	2	39	F	870	890	870	0.57
4	3	33	M	920	-	890	0.09
4	3	33	F	830	-	860	0.28

1Breeder Trials 1 and 2 each used three graded levels of cumulative crude protein (CP) and metabolizable energy (ME) intakes (High (27,778 kcal ME and 1,485 g CP), Medium (26,020 kcal ME and 1,391 g CP), Low (24,242 kcal ME and 1,296 g CP)) to 21 weeks of age. Breeder Trial 3 used only the High and Low levels.

Conclusion

It is not unreasonable to assume that birds with greater genetic potential may need more nutrition in order to fully express their phenotype as the animals with greater genetic potential in this case would be expected to have a higher maintenance requirement and possibly a higher critical body weight to initiate and maintain reproduction, as suggested by these studies.

Many long held ideas about broiler breeder management apparently require thoughtful reconsideration. Nothing should be "sacred" when an analysis from a different perspective might well provide an explanation. Just because one has always done something in one manner does not mean that it necessarily has to continue to be done in that manner. One has to remember to "listen to the chickens!"

2

Broiler Breeders and Their Management

Broiler, also known as Cornish Cross, is a type of chicken raised specifically for meat production. Produced by fast-growing breeds with low mortality, broilers can be reared successfully in standard housing conditions on readily available, custom-formulated broiler feed rations.

Cross Breeds for Parent Stock (Broiler Breeders)

Consumers expect the meat from broilers to be tender and of high quality. The whole broiler production process is designed for this requirement but the same inputs are at odds with those required for egg production by broiler breeders.

The three main steps and stages in the whole broiler production process are:

- rearing and managing broiler breeders (i.e. the birds that produce eggs for hatching into broiler chicks),
- fattening of broiler chicks
- marketing and processing of finished broiler birds

The broiler producer clearly requires birds that will achieve a high body weight, with good carcass quality, over the shortest possible period of time using the minimum amount of regular feed. In addition the producer also wants birds that possess the correct body conformation, which will feather rapidly and have a minimal mortality rate. On the other hand, producers of broiler breeders (the producer of broiler parent stock) is essentially interested in all factors related to egg production and successful embryo development – onset, frequency and continuity of laying, number, size, weight, shape, and quality of eggs. This is because the producer is focused on producing as many chicks as possible for sale.

But parent birds would have been selected and bred for the fast growing characteristics they pass on to their offspring (i.e. the broiler chicks to be fattened into broilers). Birds that accumulate weight quickly in the first few weeks after hatching are generally overweight, when mature and egg production is inversely proportional to body weight. Consequently, broiler breeder hens lay only around 140 eggs per year compared with the 250 typically produced by hens laying eggs for human consumption.

A compromise must be built into the breeding programme. Failing this producers of broiler breeders are saddled with the double disadvantage of hens that lay less than 150 eggs per year and are difficult to manage because of rapid growth rate and heavy body weight at maturity. Compromise is achieved by cross breeding. Simple programmes will typically use a 'table quality' strain as the male line (e.g. Cornish) and an egg producing strain (e.g. New Hampshire) for the female line. More complicated schemes yielding a better result would be 'Cornish' males and 'New Hampshire' females, crossed with the 'White Plymouth Rock' strain. Crossbred males and crossbred females from these respective crosses are then used as broiler parent stock to breed the broiler chicks.

High selection pressure for feed efficiency and feed conversion, growth and meat/carcass quality is applied in the male strain, but much less so in the female strain. And the use of crossbred females ensures a high degree of hybrid vigour with maximum levels of egg production, egg viability and hatching success.

Broiler Growth and Management

Selection and breeding for fast growth rates in broilers form the most important processes in the world poultry industry. Male broilers achieve rapid gain from the start, and at 6 weeks of age can weigh in at 2kg (live-weight). Female birds will tend to grow at a slower rate but this has definite marketing advantages because overall consumer demand is for broiler carcasses of various weights. It is not the amount of food consumed but the efficiency of feed utilisation and food conversion into body tissue which underpins the growth rate.

Broiler producers tend to plump for white feathered strains because they result in a 'cleaner-looking' carcass after processing. But there are instances where production management considerations outweigh this and coloured-feathered strains are preferred. Examples include broiler production in countries with high rainfall and the indigenous soil is red. In these situations, red/brown Rhode Island Reds may be the most sensible choice. Feather cover must be good to maintain insulation and restrict heat loss from the body, as well as minimising incidence of skin blistering which ruins marketability of processed birds.

Many modern strains of broiler will produce yellow fat because they have been custom-bred for the American market. In markets where yellow fat is undesirable, producers should remove carotene and carotenoid pigments (coloured chemicals) from the ration. Similarly, factors that determine carcass quality in one country may not suit another. For instance, consumers in some countries may consider the body conformation, texture and taste of carcasses high quality by 'Western' standards to offer an unattractive and insufficiently chewy eating experience. For supermarket sales in general, breast meat should be broad and deep. Many such problems are overcome by incorporating local strains into cross breeding programmes to produce appropriate broiler parent stock.

Feeding Broiler Breeders (Parent Stock)

Producers of broiler parent stock (broiler breeders) have the sole aim of obtaining the maximum number good-quality, fertile eggs and hatched broiler chicks. Parent stock will clearly possess the fast growing traits required by broiler birds, but these parent birds gain weight too quickly and hens suffer reduced egg production. Clearly there is an obvious conflict of interests to ensure proper growth and development with maximum egg production. This problem is overcome by carefully planned feed restriction using the following guidelines:

- Producers must ensure broiler breeder birds attain their mature weight at 24 weeks of age
- Give crushed millet or maize once per day during the rearing period and increase to three times a day during the laying period. Throw cereal on to the litter. Birds scramble for the grains, exercise and burn off fat

- Reduce feed by 5 g a day, for every 50 g a bird registers over the optimum weight
- If rearing males and females separately, provide males with 30g of extra feed per day. If reared together only increase the feed by 5 g a day for each male bird.

Broiler Breeder Management

Broiler breeder production employs a system much like that used to rear laying bird chicks. Use the same vaccination programme plus administration of an avian encephalomyelitis vaccine in the drinking water when birds are 18 weeks old. Cull low-quality chicks (usually 3-5 per cent) at 6 weeks and use the same pattern of lighting offered to layer birds. Immature broiler breeder birds eat excessively to satisfy their inherent (custom-bred) fast growth rates. As a result they grow too fast and become grossly overweight. Compensatory feed restriction techniques, including reduction in daily ration, low protein, high-fibre diet; miss a day feeding, restricting time access to feed and low lysine levels, are required to alleviate the problem.

On balance it is best to rear cockerels and hens separately at first because they have differing nutrient requirements. For instance, cockerels require higher inclusions of calcium and phosphorous. The sexes can then be mixed at 12 to 14 weeks using a ratio of 8-10 hens to one cockerel. All pullets and cockerels should be re-housed in laying quarters at the same time when 21 weeks of age. And with a lighting regime as for layers, broiler breeders should attain full egg production between 30 and 33 weeks of age.

Specific problems may arise in the tropics because the inherently heavier broiler breeders are more susceptible to heat stress than are birds from standard laying strains. For health considerations laying houses should be well away from houses where immature birds are being reared. Once vacated, houses used for rearing immature birds must be thoroughly cleaned and left for at least three weeks before introducing new chicks.

Turkey Rearing and Feeding Program

Turkey Introduction

Turkey poults have been described as suicides looking for somewhere to happen, especially day-old poults. More than any other species of poultry, getting them through the first days of their lives is a challenge to producers. Today's turkeys have a tremendous potential for growth; the aim of the grower should be to provide an environment which, through thorough cleaning and preparation of the house, correct preheating, proper lighting and provision of adequate feeders and drinkers, will encourage efficient consumption of feed and allow the poults to express their growth potential.

Clean Out & Disinfection/Sanitation

Prior to placement, the brooder house and all equipment to be used should be thoroughly cleaned and sanitized.

Setting up the Brooder House & Placing the Poults

A. Brooder Rings: After the house has been cleaned and dried, and 3-5 inches of new litter is spread, brooder rings should be set up. These serve several functions:

   1. To prevent drafts.
   2. To prevent pile-ups of poults.
   3. To keep the poults near food water and heat.

   Prior to poult placement, the temperature directly under the brooder should be 95 - 100 degrees Fahrenheit (35 - 38 degrees Celsius), while the temperature at the perimeter of the ring should be 90 - 95 degrees Fahrenheit (24 - 30 degrees Celsius).

   Once the poults arrive, use their behaviour to adjust the brooder temperature and height.

   Between 3 and 5 Days after placement, gradually extend the brooder guards/rings. Between 6 and 14 days, reduce brooder temperature approximately 5 degrees Fahrenheit (3 degrees Celsius) per week and gradually raise the height of the brooder. If the temperature is correct, the poults will locate feed and water sooner, which will help prevent dehydration and mortality.

B. Feed: Feeder space recommendation is 1.5 linear inches (4 cm) per poult. Feed troughs or pans should be no more than 1.5 inches (4 cm) above the litter. For the first 2 to 3 days, feed provided in egg flats or meat trays improve feed consumption. Replace these gradually with feed troughs. Feed troughs and trays should be kept as clean and as full as possible, and should be distributed so that poults can eat at whatever temperature is comfortable to them.

C. Water: Provide 0.5 linear (1.25 cm) per bird. Keep in mind the smallest bird when placing waterers and feeders. Birds will consume about 2 to 2.5 times as much water as feed.

D. Water Quality Guidelines: Poults, above all other species, are susceptible to high sodium levels in the water. Maximum recommended should be 300 - 350 ppm; at higher levels than this, the sodium level in the diet needs to be adjusted.

E. Waterers and Feeders: Waterers and feeders should be cleaned daily for the first week to keep them as fresh as possible. Water drinkers may require daily cleaning (emptying and scrubbing) for a longer period, depending on water quality.

F. Lighting: On days 1 and 2, poults should receive 24 hours of light with intensity of 60 - 70 lux (6 - 7 foot candles) at poult eye-level. This high intensity during the first two days will keep poults active, eating and drinking.

The First Weeks/Grow-out

A. Air Temperature: The goal is to drop temperatures about 2 degrees Celsius per week, getting to 22 degrees Celsius (75 degrees Fahrenheit) at the end of the fifth week.

B. Ventilation: A successful ventilation program should:

1. Supply adequate oxygen necessary for birds' respiration.
2. Maintain air temperature at a level which will be comfortable for the birds, while not wasting energy (fuel and/or feed) doing so.
3. Remove moisture from the house and maintain good litter conditions.
4. 4. Minimize dust and air-borne contaminants.

B. Grit: The feeding of insoluble grit aids the bird's digestion. The following chart is a guideline for providing grit to turkeys.

Age in Weeks	*Amount/1000 birds/day*	*Size of Grit*
0 - 5	5 kg (11 lbs)	Starter (#1)
6 - 12	7 kg (15 lbs)	Grower (#2)
13 - market	11 kg (24 lbs)	Adult (#3,4)

Feeding Ducks

Ducks are raised as pets on small ponds or lakes, for release in hunting preserves or conservation areas, and for eating purposes. The mallard is the most popular duck breed in the United States. Domestic ducks, such as the White Pekin and the Muscovy, are also popular. The commercial duck industry in the United states relies primarily on the White Pekin for meat production.

A combination of good nutrition and proper management are essential for raising healthy ducks. Maximum efficiency for growth and reproduction can be obtained by using commercially prepared diets. Because pet ducks are generally raised on open ponds or lakes, they are subject to predation. Predators that can damage your duck flock include: turtles, owl, hawks, raccoons, skunks, opossums, cats, and dogs. If possible, your ducks should be maintained in an enclosure that prevents predator access. If predation becomes a problem, recognition of the predator is imperative. Contact the state Conservation Department or Wildlife Resource Commission on the methods and legalities of removing predators from your property.

Feed Quality

Good commercially prepared duck feed, is available from most local feed stores. Some large duck operations may find mixing the complete feed on the farm to be less expensive than purchasing it from a commercial source. Regardless of whether feed is purchased or mixed on the farm, it must be stored away from rodents and insects in a clean, dry place to prevent contamination and mold growth. A pair of rats can eat or contaminate over 100 lbs of feed in a year. Use the feed within 3 weeks of the manufacturer's date and sooner during hot, humid weather to prevent loss of vitamins and mold formation. Stale or bad-smelling feed is evidence of spoilage and possible mold contamination. Never use feed that is moldy because some molds produce toxins which could cause serious health problems or poor growth. Ducks are extremely sensitive to mold toxins. For example, ducks are sensitive to as little as 30 of ppb aflatoxin. Mold toxins can cause damage to the ducks' digestive organs, liver, kidneys, muscles, and plumage, and can also reduce growth and/or reproductive performance.

The quality of feed ingredients is also very important. Do not use grains that are contaminated with molds, weed seeds, or dirt. Avoid using old vitamin/mineral packs because they lose their effectiveness with time, especially if they are exposed to sunlight or heat.

Feed Form

High quality pelleted feed is important to maximize the growth rate and feed efficiency of ducks. Performance will decrease as the amount of fines in a pelleted feed increases. Commercial pellet binders are often used to limit fines and improve pellet integrity. Although ducks can be fed mash feed, growth performance will be reduced by about 10% in comparison to that of ducks fed pelleted feed and feed wastage will be increased.

Ducklings should be fed a starter diet from hatch to 2 weeks of age. The starter diet should be fed as 1/8 inch (3.18 mm) diametre pellets or as crumbles. After 2 weeks of age, feed a grower diet as 3/16 inch (4.76 mm) diametre pellets.

Feeders and Waterers

Growing ducks should be allowed free access to feed and water at all times. Proper feeder and waterer height, maintenance and sanitation are essential for achieving uniform flock growth and health. Small feeders should be used until the ducklings are 2 weeks of age. Larger feed hoppers should be used for older ducks. The feeder pan height should be at a level even with the back of the average duck. Waterer pan height should be even with the lower neck area of the average duck and water nipples should be adjusted at a slightly higher level. Feeders and waterers that are too low result in excessive wastage. Those that are too high restrict feed and water access to the smallest ducks and thus increase size variation in the flock.

Waterers and feeders must be kept as clean as possible at all times. Shelter feeders from the sun, wind, rain, and snow to minimize feed spoilage. Feed hoppers that are used outdoors should have lids that fit securely. If feed hoppers are placed within a building or pen and water supplies are placed outside, the hoppers should be closed overnight to prevent the ducks from choking on dry feed. Water may be supplied in hand-filled water fountains or by automatic waterers.

To prevent wet litter, place the water supply above wire flooring or on a screened drain when in confinement. Waterers should be cleaned and sanitized with a commercial non-toxic disinfectant at least 3 times a week. Avoid pouring the rinse water on the litter, rather pour it into a bucket and remove it from the pen to help maintain a dry, clean environment for the ducks. Check daily to see that the waterers and feeders are working properly and not leaking or spilling.

Because young ducklings grow rapidly, they should have adequate floor, feeder, and waterer space. For the first three weeks, allow 2 square foot of space per duckling on wire and 1 square foot per duckling on litter. If confinement rearing is practiced, increase the floor space to 2.5 square feet per duckling through 7 weeks of age. Ducks should be given at least 1.5 linear inches of feeder space and 0.5 linear inches of waterer space per duckling at all times. Larger ducks such as the Muscovy may require some additional space. Ducks are waterfowl, so they are instinctively attracted to water. This characteristic can cause a serious wet litter problem if the waterer is not designed properly or maintained at a proper height. If ducks are raised in confinement and subsequently released to a pond, a water bath is helpful for ducks to preen an keep their plumage properly oiled. This will help keep their feathers in good condition and give them the ability to swim in a pond.

Nutritional Requirements

Dietary factors affecting growth and feed utilisation have been established for several varieties of ducks. Suggested macro- and micro-nutrient requirements of most ducks in different age categories are listed in table, respectively. Example rations that follow these nutrient specifications are listed in table below.

Table: *Suggested Macronutrient Requirements of Ducks*

	Starter	*Grower*	*Finisher*	*Breeder*	
Nutrient	*0-2 weeks*	*2-6 weeks*	*6-8 weeks*	*Developer*	*Layer*
Metabolizable energy (Kcal/lb.)	1400	1400	1400	1175	1300
% Protein	20.0	18.0	16.0	14.5	16.0
% Lysine	1.1	0.9	0.8	0.65	0.75
% Arginine	1.1	1.0	0.9	0.7	0.85
% Methionine + Cystine	0.9	0.8	0.7	0.6	0.65
% Calcium	0.9	0.8	0.8	0.7	2.9
% Available Phosphorus	0.45	0.4	0.4	0.35	0.35
% Linoleic Acid	1.0	1.0	1.0	0.8	1.0

1. Nutrients shown in this table apply only to the energy level specified.
2. Begin feeding breeder layer feed one month before the first egg is laid.
3. The energy concentration given is only an example. The energy concentration may vary from 1000 to 7500 Kcal/lb, provided the concentration of each nutrient per unit of energy remains the same.

Table: *Suggested Micronutrient Requirements of Ducks*

Nutrient	*Vitamin-Mineral Level*		
	1 *0-2wks*	*2* *2wks-adult*	*3* *Breeder*
Minerals			
% Potassium[2]	0.7	0.6	0.6
% Sodium	0.17	0.14	0.14
% Chlorine	0.12	0.12	0.12
Magnesium (mg/lb.)	230	230	230
Manganese (mg/lb.)	25.0	25.0	25.0
Zinc (mg/lb.)	32.0	25.0	30.0
Iron (mg/lb.)	35.0	20.0	30.0
Copper (mg/lb.)	4.0	3.0	3.0
Iodine (mg/lb.)	0.18	0.14	0.20
Cobalt (mcg/lb.)	90.0	90.0	90.0
Selenium (mcg/lb.)	70.0	70.0	70.0
Vitamins			
Vitamin A (IU/lb.)	4000	2500	4000
Vitamin D3 (ICU/lb.)	500	400	400
Vitamin E (IU/lb.)	10.0	5.0	10.0
Vitamin K (mg/lb.)	1.0	0.5	1.0
Riboflavin (mg/lb.)	3.0	1.5	3.0
D-Pantothenic acid (mg/lb.)	6.0	4.0	5.0
Niacin (mg/lb.)	25.0	20.0	25.0
Vitamin B12 (mcg/lb.)	4.0	2.0	4.0
Choline (mg/lb.)	900	450	450
Biotin (mg/lb.)	0.05	0.05	0.05
Folic Acid (mg/lb.)	0.60	0.40	0.50
Thiamin (mg.lb.)	1.6	1.5	1.4
Pyridoxine (mg/lb.)	1.4	1.4	1.4
Ethoxyquin (mg/lb.)	60.0	60.0	60.0

1. Vitamin-Mineral Level(s) should provide the following levels/ pound of complete feed.
2. Not needed in commercial Vitamin-Mineral premixes.

Table: *Example Rations for Ducks*

	Percentage of Complete Ration				
	Starter	*Grower*	*Finisher*	*Developer*	*Layer 1*
Ingredient					
Yellow corn, #2 dent	70.00	73.58	77.25	39.50	59.00
Barley	--	--	--	15.00	15.04
Oats	--	--	--	11.20	--
Soybean meal (48% protein)	18.18	19.70	16.13	12.40	13.95
Alfalfa meal (17% protein)	2.00	--	--	--	--
Fish meal (60% protein)	7.50	--	--	--	--
Meat & Bone meal (50% protein)	--	5.00	5.00	--	5.00
Wheat Bran	--	--	--	10.00	--
Wheat Middlings	--	--	--	8.00	--
D,L-Methionine	0.17	0.22	0.16	0.15	0.14
Dicalcium Phosphate (18.5% protein)	0.55	0.28	0.15	1.30	0.18
Ground Limestone	0.75	0.77	0.86	2.00	6.24
Iodized salt	0.25	0.25	0.25	0.25	0.25
Vitamin-mineral package	0.202	0.203	0.203	0.203	0.204
Chlortetracycline-50	0.40	--	--	--	--
Calculated Analysis					
% Protein	20.00	18.30	17.00	15.00	16.00
Metabolizable energy (Kcal/lb.)	1400	1410	1426	1200	1312
% Calcium	0.90	0.85	0.80	0.75	2.90
% Available Phosphorus	0.45	0.40	0.35	0.38	0.35
% Lysine	1.12	0.90	0.80	0.70	0.75
% Methionine + Cystine	0.90	0.80	0.70	0.65	0.65

1. Layer diet may be supplemented with free choice ground oyster shell.
2. Supplies/pound of complete ration the vitamins and minerals in the amounts listed on package No. 1 (Table 2).
3. Supplies/pound of complete ration the vitamins and minerals in the amounts listed on package No. 2 (Table 2).
4. Supplies/pound of complete ration the vitamins and minerals in the amounts listed on package No. 3 (Table 2).

Small Flock Ducks

Feed that is especially prepared for ducks is ideal. Beware of cheap rations that contain a lot of by-product ingredients because they may cost you money in the form of decreased body weight gain, poor feathering, reduced egg production, and hatchability, or other problems. A quality feed from a reputable dealer is usually the most profitable feed in the long run. If availability or the cost of duck feed is a major limitation, chicken feed could be used as an alternative. A 23% protein chick starter could be used for the first 2 weeks, followed by a 20% protein broiler grower diet. If available, a broiler finisher diet containing 18% protein can also be used. Be cautious, however, because broiler chicken feed may contain feed medications that do not have the Food and Drug Administration's approval for ducks.

Feed Medications to Control Disease

Ducks exhibit greater resistance to most diseases and parasites than do most domestic fowl. As a consequence, medicated feeds for ducks are used less often than with chickens and turkeys. Presently, there are feed medications available to control the common diseases of ducks: colibacillosis, fowl cholera, salmonellosis, and necrotic enteritis.

Colibacillosis is a common disease in ducks caused by the bacterium Escherichia coli. E. Coli can cause embryonic and duckling mortality by infecting the yolk sac. Infection of the digestive track and air sacs is most common. Infected ducklings appear droopy and listless and exhibit diarrhea and ocular discharge. Cleanliness of the hatching eggs and good management in the hatchery are necessary for prevention of Colibacillosis. The combination of sulfadimethoxine at 0.05% of the diet and ormetoprim at 0.03% of the diet for a duration of 7 days can reduce or prevent mortality from Colibacillosis in baby ducklings. Fowl cholera is a contagious disease of domestic ducks and other birds, caused by the bacterium Pasteurella multocida. Sick ducklings refuse feed and exhibit diarrhea and mucus discharge from the mount.

Mortality may be as high as 50%. A concentration of 0.44% chlortetracycline (400 g/ton) in feed is effective in reducing mortality. Treat infected ducks for 5 days. Chlortetracycline binds to calcium in breeder feed, thus a low calcium diet (0.6-0.8%) should be used during the 5 day treatment period.

Salmonellosis is a common disease of ducks caused by a variety of serotypes of salmonella. Infected ducks are listless, dehydrated, exhibit diarrhea, and show signs of incoordination, and head tremors. Mortality is about 10%.

Salmonellosis can be treated with chlortetracycline (.044%) or sulfadimethoxine-ormetoprim (0.04-0.08%) in the feed. Necrotic enteritis is a common infectious disease of breeder ducks. The exact cause is not known. Infected ducks are weak and unable to stand, and their digestive tracts are swollen and filled with blood-stained fluids. Mortality is high, approximately 40%. A concentration of 0.02% neomycin sulfate in feed for 2-3 weeks can reduce mortality.

The effectiveness of a disease prevention program, regardless of feed medication usage, is best under good management and sanitary practices. Try to keep houses clean and dry, and do not allow mud holes and slimy areas to form. Always consult a veterinarian for proper medication usage or if a disease problem is persistent or serious.

Minimum Nutritional Requirements of Geese

Table 1 gives the minimum nutritional requirements of geese. Table 2 and Table 4 give suggested rations for goslings and breeding stock.

Table: *Minimum basic nutritional requirements of geese*

Nutrient	*Grower*	*Breeder*
Protein, starter	20%	16%
Protein, finisher	16%	–
Energy	11 100 kJ/kg	10 500 kJ/kg
Fibre	4%	5%
Fat	5%	4%
Calcium	1%	3%
Methionine	0.30%	0.25%
Phosphorus	0.60%	0.50%
Vitamin A, as retinyl (acetate)	3100 mg/kg	41 300 mg/kg
Vitamin D3	300 mg/kg	65 mg/kg
Riboflavin	10 mg/kg	10 mg/kg
Calcium pantothenate	15 mg/kg	20 mg/kg
Niacin	55 mg/kg	55 mg/kg
Manganese	60 mg/kg	40 mg/kg

Feeding Goslings

Goslings are often fed a similar ration to ducks, but because goslings show a rapid weight gain during the first 4 weeks, they need more protein. The heavy breeds of geese weigh approximately 85–100 g at day-old and may weigh up to 1.6 kg at 4 weeks of age.

Provided there is plenty of green feed, goslings can begin to graze at just a few weeks of age.

- The liveweight of geese will increase by up to 50% during their first 2 months of life.
- Goslings grow more rapidly when housed and fed a completely prepared and well-balanced ration than when they graze. To 10 weeks of age, goslings reared in cages will weigh up to 20% more than floor-reared goslings. Day-old birth weight and rate of growth in the first month influence a gosling's weight at 10 weeks.
- A starter diet containing 20% protein is recommended for the first 4 weeks in conjunction with good grazing. After 4 weeks, feed goslings a finisher ration containing 16% protein.
- Starter and finisher rations may be fed either wet or dry, in mash or pelleted form.

Table 2 gives suggested rations for goslings.

Table: *Suggested rations for goslings (a vitamin and mineral premix should be added to these rations)*

Ingredients	*Starter (%)*	*Finisher (%)*
Wheatmeal	34.75	40.75
Sorghum meal	20.00	30.00
Bran	10.00	6.00
Pollard	8.00	6.00
Coconut meal	–	–
Meatmeal	18.00	12.00
Lucerne meal	5.00	3.00
Milk powder	4.00	2.00
Ground limestone	–	–
Salt	0.25	0.25
Total	**100.00**	**100.00**

Restricted Feeding Program

Where good grazing is available, practise supplementary feed restriction when the finisher ration is fed. Allow 500 g feed per head per week for goslings between the ages of 4 and 8 weeks, then 1 kg feed per head per week to 12 weeks of age. At the end of the restricted feeding program, allow goslings an unrestricted ration to marketing.

The type of restricted feeding program adopted depends on many factors, including the required gosling weight for marketing, the amount of pasture available, the quality of the pasture and the time of year. Suggested programs for restricted feeding, together with a guide to expected body weights, are given in Table 3.

Table: *An Indication of Expected Performance from a Restricted Feeding Program*

Program	*Age (weeks)*	*Bodyweight (kg)*
Complete ad lib feeding to marketing.	10	>4.0
Full complete feeding to 4 weeks of age. Pasture plus 130 g of complete feed to 12 weeks of age. Full complete feeding to marketing.	15	5.0
Full complete feeding to 3 weeks of age. Pasture only to 18 weeks of age. Full complete feeding to marketing.	21	6.5

Feed Conversion

Feed conversion is calculated by dividing feed consumption by the bird's live bodyweight. In other words, it is the amount of feed eaten (in kilograms) required to produce 1 kg of meat (liveweight). For example, goslings fed without restriction on a balanced ration to marketing at 10 weeks will have a feed conversion of approximately 3:1. This means a total of 12 kg of feed will be consumed for a bodyweight gain of 4 kg. As goslings get older, feed conversion capability diminishes.

Growth Promotants

The inclusion of growth promotants in a gosling's diet will improve the rate of growth. However, since geese are sensitive to arsenicals, this additive must not be included in the ration.

Note: Force feeding for any purpose is unacceptable in Australia.

Feeding Breeders

Good grazing for all breeding stock is needed for up to 6 weeks before the breeding season. Then, up to and during the breeding season, feed geese a ration with about 16% protein. Rations for laying hens are suitable.

Breeding geese in full lay should be given about 200 g of prepared feed a day, depending on the amount of pasture or green feed available.

Geese should have access to both soluble and insoluble grit at all times. Soluble grit is in the form of limestone chips (5 mm) or shell grit, whilst insoluble grit is usually supplied as blue metal or basalt chips screened to 5–6 mm.

Table: *Suggested rations for breeding stock (a vitamin and mineral premix should be added to these rations)*

Ingredients	*Ration 1 (%)*	*Ration 2 (%)*
Wheatmeal	30.75	57.75
Sorghum meal	20.00	22.00
Bran	12.00	–
Pollard	12.00	–
Coconut meal	8.00	–
Meatmeal	10.00	13.00
Lucerne meal	5.00	5.00
Milk powder	–	–
Ground limestone	2.00	2.00
Salt	0.25	0.25
Total	**100.00**	**100.00**

Grazing Pasture

Geese are more like grazing animals than any other type of poultry. Their beak and tongue are particularly well-equipped for grazing. The beak has sharp interlocking serrated edges designed to easily cut and divide grass and other plant tissue. The tongue at the tip is covered with hard, hair-like projections, pointing towards the throat, which quickly convey the pieces of grass and other vegetable material into the throat. This rough covering on the point of the tongue enables geese to bite off plants even closer to the ground than sheep can. Because of this, overstocking must be avoided as the ground will become bare. Because geese have virtually no crop in which to hold feed, they tend to feed and graze frequently. In summer they may continue to graze and feed at night.

As stated, goslings can start grazing at just a few weeks of age. If pasture is good and plentiful then the amount of prepared food they are given can be reduced. A system of rotational grazing should be practised to ensure geese have access to good pasture all the time. Where paddocks are fenced off and allowed to spell, the pasture will regrow quickly and the paddocks will be more hygienic. Overseas research has shown overall performance of breeding geese is greatly improved when they have access to good pasture.

If pasture for grazing is not available then breeders should be fed chopped green feed. Geese prefer to pick their own green feed and may reject cut grass unless it is fresh and very finely chopped. Geese can be very selective in the pasture they eat and tend to pick out the more palatable pastures. They reject narrow-leaved tough grasses and select the more succulent clover and grasses. The stocking density for geese on pasture will vary depending on the quality of the pasture and the age and size of the geese. But as a guide, growing geese can be stocked at a density of 50–100 birds/ha, and breeding geese at about 20 birds/ha. During the non-breeding season, breeding geese only need access to pasture to fulfil their total feed requirements.

Feeding Game Birds

The following topics are discussed in this information sheet:

Feeding Programs

Vitamins

Minerals

Medicated Feeds

Dietary Formulations

Water

Nutritional Assistance

All poultry and game bird feeds are referred to as "complete" feeds. They are designed to contain all the protein, energy, vitamins, minerals, and other nutrients necessary for proper growth, egg production and health of the birds. Feeding any other ingredients, either mixed with the feed or fed separately, will upset the balance of nutrients in the complete feed. Feeding additional grain or supplement with the complete feed is not recommended.

Young game birds kept for meat production are fed differently than birds saved for egg production or breeding. In addition, meat-type Bobwhite quail have larger bodies and are expected to gain weight more rapidly than birds grown for "flight" purposes. Therefore, birds are fed diets that contain nutrient levels that reflect the dietary needs of the specific type of birds being produced. Meat-type birds are grown as flight birds will be more expensive to produce since they will consume more feed, be larger than necessary and are not considered as good fliers. In contrast, smaller strains of Bobwhite quail that are usually considered as good flight birds are not recommended as good meat producers. They are not efficient converters of feed to meat and produce less desirable carcasses when slaughtered.

Feeding Programs

Feed game bird chicks a "starter" diet soon after hatching. Continue feeding the starter until they reach six or eight weeks of age. The starter diet has the highest level of protein that a bird receives during its lifetime. As the chicks age they require lower levels of most nutrients including dietary protein but need a higher level of energy. After the chicks reach six or eight weeks of age, feed them either a "finisher" diet (meat-type birds) or a "developer" diet (flight birds or those saved for egg production). Feed meat birds a finisher diet until they reach slaughter size. Feed the flight birds and immature breeders the developer diet until they are sold or about twenty weeks of age. A few weeks prior to expected egg production, the breeders are fed a "layer" diet until they complete their egg production period.

An alternate species of game birds often produced are the coturnix or pharaoh quail. They are grown for both meat and egg production but seldom for flight or hunting. They mature at an earlier age than bobwhite quail and may begin laying eggs as young as six to eight weeks of age. As with bobwhite quail, coturnix grown for meat are provided starter and finisher diets, whereas laying/breeder birds are fed starter and breeder diets.

The minimum dietary requirements for protein, calcium and phosphorus for game bird feeds are shown in Table 1. It is important to provide the correct diet to the birds if desired result are to be attained. Remember, breeders saved for egg production are fed developer diets, not finisher diets. Laying/breeder birds are fed only laying diets. Otherwise, you will observe reduced egg production and increased numbers of thin-shelled eggs.

Vitamins

Vitamins are always added to feeds in amounts that meet minimum dietary requirements. This insures that birds consume plenty of vitamins for proper health and performance. Higher levels are not usually harmful, but extra vitamins are unnecessary and expensive. Minimum vitamin requirements for various ages of birds are shown in Table 2.

When adding vitamins to the diet as a premix, make sure that an adequate amount of all vitamins are provided. It may be necessary to add extra amounts of some vitamins to achieve minimum levels for other vitamins. This may increase the cost of the complete feed but is better than creating vitamin deficiencies that can be more expensive. During periods of stress caused by disease, shipping or sudden changes in the environment it is recommended that extra vitamins and electrolytes be provided in the drinking water until the stressing condition is corrected.

Minerals

Like vitamins, adequate levels of minerals must be provided to all birds. Minerals in breeder feeds are especially important. Laying hens require higher levels of minerals for egg shell formation. Chicks require high levels of minerals for proper bone formation and development. Breeder feeds are fed only to laying birds. If a breeder feed is fed to chicks reduced growth and unnecessary stress will be placed on the chicks. Although not always required for survival, better performance is observed if a trace mineral premix is added to diets. Trace minerals are those minerals required at very low levels for good growth and production. Most feed ingredients provide some of these minerals but sometimes contain less than adequate quantities. Many of these minerals are contained in commercial vitamin premixes. An excellent trace mineral premix is shown in Table 3. The premix will provide adequate trace minerals when added at the rate of two pounds per ton of feed.

Medicated Feeds

Game bird feeds are available with several types of medications for preventing or treating diseases. The two most common medications added to feeds are coccidiostats and/or antibiotics. Coccidiosis is a parasitic disease of the digestive tract. It is difficult to control by sanitation practices alone. The best prevention is to include a drug or coccidiostat in the feed. This coccidiostat is added to the feed at low levels and fed continuously. Some coccidiostats can be given at higher levels to treat the disease after the birds show symptoms. Before increasing the drug level, check with someone who is familiar with the proper use of the coccidiostat in question since some coccidiostats can be toxic at higher levels.

Feed growing birds a ration containing a coccidiostat from hatch until the last week before slaughter. Feed an unmedicated diet during this last week to assure that no drug residues remain in the tissues of the birds. This feeding of unmedicated diets prior to slaughter is recommended when using any dietary drug, regardless of whether the restriction is required or not.

As birds mature, they develop a resistance to coccidiosis if a controlled exposure to the parasite is allowed. Birds grown for breeder replacements are fed a coccidiostat until about 16 weeks of age. The medicated feed is then replaced with a feed not containing a coccidiostat. Spotty outbreaks of the disease can be controlled by including drugs in the water. Two coccidiostats with Food and Drug Administration (FDA) approval for use in game bird feeds are monensin sodium (Coban) and amprolium. Antibiotics may also be added to some feeds. Antibiotics aid performance and maintain healthy birds. They are added at low (prophylactic) levels to prevent minor diseases and produce faster, more efficient growth. Higher (therapeutic) levels are usually given in water or injected into the bird. Examples of FDA approved antibiotics in game bird diets are bacitracin and penicillin.

Addition of bacitracin to game bird diets is recommended at the rate of 50 grams per ton as a preventative of ulcerative enteritis (quail disease). Higher levels in the diet is not recommended nor permitted by FDA. If higher levels are needed for treatment, it is best to give the antibiotic in the birds' drinking water. This practice is also more effective since sick birds will usually drink water but will not always consume feed. Including bacitracin in diets of all game birds is recommended to maintain healthy, productive birds.

When using any drug, whether the drug is or is not mixed with the feed, all warnings and instructions on the label must be carefully followed. Always comply with all instructions that require a medication withdrawal period prior to bird slaughter or saving eggs for human consumption.

Diet Formulations

Several diets are included that provide adequate levels of all nutrients for the type of birds cited. Diets for meat-type Bobwhite quail are presented in Table 4, flight-conditioned Bobwhite quail in Table 5, and Coturnix or pharaoh quail in Table 6. All ingredients must be used without substitution or alteration of quantities if satisfactory results are expected. Any deviation from the recommended diet will alter the levels of all nutrients and possibly create undesired problems. Always consult with a poultry nutritionist or your County Agent before making dietary changes.

Most commercially prepared game bird feeds are in "crumble" form. These feed particles are formed by pelleting the mixed "mash" feed followed by partial regrinding or crumbling into the desired size particles. Often the crumbles of starter feeds are too large for newly hatched quail to eat. Therefore, it is necessary for additional grinding until the particles are the desired size. Crumbles are not necessary for good production although they have several desirable characteristics. Mash diets made from the dietary formulations shown in Table 4 will produce excellent performance. The assortment of ingredients used has intentionally been kept to a minimum. Many additional ingredients can be used, but the diets need reformulation to compensate for nutritional differences. Attention to high quality ingredients is required when making bird feeds.

Prior to the start of feed manufacturing, make sure that all ingredients are available. High quality ingredients are mandatory if satisfactory results care expected. Often poor quality ingredients are used when making diets for other types of farm animals and poor performance is not observed. If these same ingredients are used in game bird feeds, it is assured that you will experience production problems. Never use a feed ingredient unless it is of highest quality.

Often high-quality commercial game bird feeds are not available and substitutes are needed. Comparable turkey feeds can be substituted for game bird feeds without reduction in performance. In most cases, chicken diets can be fed to growing bobwhite quail that are raised for slaughter.

Check with a qualified nutritionist before making dietary substitutions. If production problems occur that are thought to be "feed related", the first action is to obtain a sample of the suspected feed. About one quart of the feed is adequate in most situations. Contact an Extension Poultry Specialist and obtain his/her assistance in solving the problem. A one to two cup portion of the feed sample is usually submitted to a feed analysis laboratory to check for undesirable nutritional characteristics. Store the remaining sample for future reference.

If production problems are unusually severe, temporary replacement of the suspect feed may be necessary until the cause for the problems is determined. Only use a suitable diet from another feed manufacturer, and preferably, from a different feed dealer. Obtaining additional feed from the same dealer and manufacturer may result in bringing more of the same cause for your problems. If, after determining the cause of the problem and it is not feed related, you can resume use of you favourite dealer's feed.

Water

Many producers overlook the importance of providing clean, fresh water to their flocks. Water, though not considered a nutrient by many producers, is the most important nutrient for animals. Like all farm animals, game birds need clean water at all times. Drinking water must not get too hot or cold, or the birds will not drink it. Clean the water troughs and replace with fresh water at least once daily. You must keep water and feed troughs clean from droppings, litter, soil and other contaminants.

Keep feed troughs clean and dry. Place them so the feed stays dry. Empty the feed troughs at least two or three times weekly (daily if necessary) and refill with dry, fresh feed.

Do not wash feed troughs unless they are excessively contaminated with wastes or unless the feed gets wet. Do not let the feed get moldy. Moldy feeds can kill quail.

Nutritional Assistance

If you need assistance with any problem related to game bird production, contact your local County Agent or the Extension Poultry Specialists at Mississippi State University.

Table: *Recommended Nutritional Requirements*

Diet	*Protein(%)*	*Calcium(%)*	*Phosphorus(%)*
Bobwhite Quail			
Meat-type			
Starter (0 - 6 wk)	23.0	1.0	.50
Finisher (6 wk - mkt)	19.0	90	50
Flight			
Starter (0 - 6 wk)	26.0	1.0	50
Developer (6 - 16 wk)	22.0	.90	.50
Breeders			
Starter (0 - 6 wk)	26.0	1.0	.50
Developer (6 - 20 wk)	18.0	1.0	.50
Layer (20 wk +)	19.0	2.75	.65
Coturnix (Pharaoh) Quail			
Starter (0 - 6 wk)	24.0	.85	.60
Finisher (6 wk - mkt)	18.0	.65	.50
Layer (6 wk +)	18.0	2.75	.65

Table: *Vitamin Requirements for Finished Feeds*

	Amount of vitamin per:	
Vitamin (units)	Pound	Ton
Vitamin A (IU)	2000	4 million
Vitamin E (IU)	15	30,000
Vitamin D3 (IU)	1000	2 million
Vitamin K (mg)	.3	600
Riboflavin (mg)	2	4,000
Pantothenate (mg)	8	16,000
Niacin (mg)	20	40,000
Pyridoxine (mg)	2	4,000
Vitamin B12 (mg)	.005	10
Choline (g)	.7	1,400
IU = International Unit, mg = milligram, g = gram		

Table: *Trace Mineral Premix*

Ingredient	Amount for 10 lb of Premix
Manganous Sulfate	1.25 lb
Zinc Oxide	1.25 lb
Ferrous Sulfate	5 oz
Copper Sulfate	1 oz
Limestone or Oyster Shell	7.1 lb
Add to diet at rate of .1% or 2 lb/Ton.	

Table: *Ingredient Composition of Meat-type Bobwhite Quail Diets (Expressed in table as percentages)**

Ingredient	***Starter***	***Finisher***	***Breeder***
Yellow Corn	58.00	67.93	60.89
48% Soybean Meal	37.23	27.30	28.50
Wheat Middlings	---	---	---
Limestone	1.84	1.87	7.17
Dicalcium PO4	1.77	1.82	1.84
Salt	.56	.57	.44
Feed Fat	---	---	.70
dl-Methionine	.15	.06	.06
Bacitracin	.05	.05	.05
Coccidiostat	.05	.05	---
Vitamin Premix	.25	.25	.25
Mineral Premix	.10	.10	.10
*To determine the amount of each ingredient necessary for one-Ton of diet, multiply each number by 20.			

Table: *Ingredient Composition of Flight-type Bobwhite Quail Diets* (Expressed in table as percentages)*

Ingredient	***Starter***	***Conditioner***	***Breeder***
Yellow Corn	43.73	42.27	60.89
48% Soybean Meal	48.27	17.98	28.50
Wheat Middlings	---	36.43	---
Limestone	1.23	1.18	7.17
Dicalcium PO4	1.72	1.59	1.84
Salt	.43	---	.44
Feed Fat	4.03	---	.70
dl-Methionine	.14	.12	.06
Bacitracin	.05	.05	.05
Coccidiostat	.05	.05	---
Vitamin Premix	.25	.25	.25
Mineral Premix	.10	.10	.10
*To determine the amount of each ingredient necessary for one-Tonof diet, multiply each number by 20.			

Ratite Nutrition and Feeding

There is limited quality research concerning the nutritional requirements of Ratites. However, some dependable guidelines have been established because of work completed in Australia and Africa. As in all diet formulations, a variety of high quality ingredients should be used to meet the nutrient recommendations of the Ratite. Using a wide variety of ingredients helps to decrease the effect of variations that are inherent in all ingredients.

Recommendations for Ratite Diets: Probably the greatest concern of the Ratite farmer is related to the protein content of the diets being fed to their birds. Some growers feel that the higher the protein the better. This is not necessarily true. Protein value is of greater importance. When the amino acids are balanced, protein content can be reduced without decreasing the quality of the feed. In fact, high levels of unbalanced proteins can be detrimental to bird growth and performance. In a worst case scenario, if an amino acid is deficient in the diet, the birds may actually consume markedly more feed without increased performance and possibly decreased performance. Another concern of Ratite growers is related to vitamin and trace mineral levels. Again, the level of individual vitamins and minerals are important but not as important as balance. The balance of the vitamins and minerals are of utmost importance. There are many interactions between many of these required micronutrients. If one particular nutrient is very high in the diet, that nutrient may actually reduce the absorption or metabolism of another nutrient. Therefore, the addition of high levels of a particular nutrient to the diet because of a report of its importance may result in more damage than good.

Recommended Vitamin and Mineral Levels for Diets		
Item	***(source)***	***Amount per ton***
Vitamin A	(vitamin A acetate)	12,000,000 I.U.
Vitamin D 3	(cholecalciferol)	3,900,000 I.C.U.
Vitamin E	(dl-alpha tocopherol acetate)	45,000 I.U.
Vitamin K	(menadione sodium bisulfite complex)	15,000 mg
Vitamin B 12	(cyanocobalamin supplement)	25 mg
Folic acid	(folic acid supplement)	2,100 mg
Riboflavin	(riboflavin supplement)	11,000 mg
Niacin	(niacin or niacinamide)	56,000 mg
d-Pantothenic acid	(d-calcium pantothenate)	21,000 mg
Pyridoxine	(pyridoxine hydrochloride)	8,000 mg
Thiamine	(thiamine mononitrate)	4,000 mg
Choline	(choline-Cl)	450 gm
d-Biotin	(d-Biotin supplement)	150 mg
Selenium	(sodium selenite)	272 mg
Manganese	(manganous oxide)	80 gm
Zinc	(zinc oxide)	80 gm
Iron	(ferrous sulfate)	45 gm
Copper	(copper sulfate)	10 gm
Iodine	(calcium iodate)	1 gm

Suggested Minimum Nutrient Compositions				
Nutrient	***Starter 0-8 wks***	***Grower 8-25 wks***	***Maintenance Over 25 wks***	***Breeder***
Met Energy (poultry)	1200	1200	1200	1150
Crude protein (%)	18.0	17.0	16.0	16.5
Fat (%)	3.0	2.5	2.5	3.5
Linoleic acid (%)	104	1.4	1.4	1.4
Lysine (%)	0.9	0.78	0.75	0.75
Methionine and	0.7	0.60	0.55	0.60
cystine (%)	1.25	1.25	1.25	2.50
Calcium (%)	0.90	0.90	0.90	0.75
Phosphorus (%)	0.68	0.65	0.65	0.52
Available Phos (%)	0.22	0.22	0.22	0.22
Sodium (%)				

Suggested Ingredients, minimums and maximums (lbs/ton)				
Ground Yellow Corn	0-800	0-800	0-800	0-600
Wheat middlings	0-450	0-600	0-400	0-650
Soy (44% CP)	0-300	0-250	0-350	0-250
Corn Gluten Meal	0-200	0-200	0-200	0-200
Barley	0-200	0-200	0-200	0-100
Oats	0-100	0-100	0-100	0-100
Wheat	0-300	0-300	0-300	0-300
Meat & Bone (50% CP)	0-150	0-150	0-100	0-100
Alfalfa meal (dehy)	0-200	0-200	0-200	0-200
Fat	0-80	0-80	0-50	0-100

Deflourinated Phosphate, Limestone, D,L Methionine (99%), L-Lysine.HCl, Salt, Vitamin, and Minerals should be added as required to meet recommendations. When the feed is manufactured, care should be taken to produce a consistent particle size. The Starter feed should be offered in the crumbled form. All other feed should be pelleted.

Feeding Your Birds

If the decision is made to change to this type of formulation, several management procedures should be followed. Always change from one type of feed to another slowly i.e., begin mixing the new diet into the diet which you have been feeding your birds. Initially, mix 1/4 new to 3/4 present diet.

After four days, mix the diets 1/2 to 1/2. After eight days, mix the diets 3/4 to 1/4 of the old diet. After two weeks of this process the new diet should totally replace the feed from which the change was made. It is very important to make a slow transition. Problems may arise if a quick change is made. For example, birds may avoid the feed, birds may develop diarrhea, or other responses may be noted. A feeding program is only as effective as the management practices followed.

Birds should be offered an amount of feed on a daily basis that they will actually consume. Forcing the birds to "clean-up" the feed on a daily basis results in the birds consuming a more balanced diet. This keeps birds from picking through the feed and excluding certain constituents from their diet. A feed that is properly pelleted, should not be a problem. Also, leftover feed will either be wasted, get wet and mold, or draw predators and rodents. None of these alternatives are very good for production. Again, management is very important in accomplishing this recommendation. The grower must monitor the consumption of the birds very closely. Do not assume that consumption of feeds used in the past will be the same as new formulations. Feed should be weighed-in in the morning. If feed remains at night, this should be removed and weighed. Feed additions the following day should be consistent with the consumption of the previous day. Growing birds may eat more in subsequent days. If the feed runs out during the day, increase the feed input by 5 to 10 percent on the following day and record the results for future reference.

Feeding Your Pet Bird

Birds come in many shapes and sizes, which means that there are many diets and many theories on how best to feed them. Diets for pet birds have progressed significantly over the past 10 years, but there is still a great deal to learn. Pellets are now readily available and have done wonders to improve pet bird diets on average, but pellets are not always the best option. What works for one type of bird may not work for another. This article will provide some of the basic dietary information to help you determine an appropriate diet for your pet bird, and will hopefully give you some ideas on how to make your bird's diet more interesting and nutritional.

Often, the key to a healthy diet is to provide a variety of items that include lots of fresh foods from the produce department. Many of the healthiest pet birds I see are ones whose owners share their meals with them. Not only does this provide a varied diet, but it can provide regular social interaction with your bird. For dogs and cats, the advice is usually not to feed human foods, but for birds, the opposite is often quite true. Plus, it's always entertaining to watch your pet deal with a plate of spaghetti noodles.

Call of the Wild

A good diet is the most important factor for maintaining a bird's health. The key to determining the best diet for any type of bird is to understand their diet in the wild. A bird that feeds primarily on fruit in the wild will do poorly on a seed based diet in captivity. It's easy to see that a finch requires a different diet than a macaw, but similar looking birds may also have different requirements. Some finches eat a diet comprised almost entirely of seed, while others may need lots of insects. Since most of us don't have a rainforest full of exotic fruits and bugs at our disposal, it is impossible to provide the same foods that our birds would find in the wild. We can approximate it, however, by feeding a variety items that are available to us from the produce department and from feed stores. Variety is often the key.

The dietary requirements of the birds we keep may be closely tied to their wild ancestry, but we also must adjust for conditions in captivity. Keep in mind that captive birds do not have to forage endlessly to obtain the food they need. Their owners, without much effort on the bird's part, provide it for them. It doesn't take much logic to guess what an endless food supply and reduced exercise might cause. Obesity is a problem with pet birds that have steady access to rich foods and spend a large part of their time in a cage. The goal is a healthy diet, low in fat, and rich in vitamins and minerals.

General Types of Feeds

Seed - Seed is the type of food that is most commonly fed to pet birds. This is true in part because it is as readily available as a hamburger and fries, but also because most birds favour seed over other items and many will eat it as long as it is available. Unfortunately, many people feed a diet comprised entirely of seed. This will ultimately lead to a malnourished bird. Among other nutrients, seed is lacking in calcium, and vitamins A and D. Many commonly fed seeds such as sunflower and safflower are also very high in fat. Seed can be a healthy part of your bird's diet, but it must be fed sparingly to be sure other items are consumed. Also be sure the seed is fresh. Seed will go stale if stored too long. To test the freshness of your seed simply soak it in water overnight then drain it. Spread it out on a damp paper towel then roll it up and let it sprout for a couple of days at room temperature. Nearly all the seed should germinate. If it doesn't, the seed is too old and stale and has lost its nutritional value. Birds that are fed on stale seed can literally starve to death with full food dishes.

Pellets - Pellets are now readily available from a number of manufactures. In general, these are all good products. Some birds will thrive on a pellet diet, but other types of birds need more than just pellets. Pellets can be fed to provide a good foundation of a healthy diet. If feeding other foods in addition to pellets, it may be necessary to limit these treats so that the bird is motivated to eat the pellets as well. One great plus to pellets is that manufacturers are developing new breed specific formulas to help with the specific dietary needs of certain types of birds. Beans and Grains – Also sold as "Sprout mixes" or "soak & cook mixes".

These usually include a mixture of beans and grains and sometimes pasta noodles. These mixes are excellent protein and vitamin sources with low fat content. Because these mixes are fed either cooked or sprouted, they are more digestible and nutritious.

Produce - Fresh produce is the area where we can really add to the variety of foods we can provide to our birds. Don't be afraid to experiment with different items since just about anything that is nutritious for people is good for birds. Apples are universally accepted, but some really love fresh yellow chili peppers. The only items to avoid are avocado and apricot pits.

The skin and pits of avocado contain potential toxins for parrots in addition to the fact that it is high in fat. Insects – Thankfully, insects are not a part of most pet bird diets so don't run out with bug jar in hand. Insects are important for many types of finches and softbills however. The easiest sources are mealworms and crickets that are commercially available from pet stores or fishing bait shops.

Summary

If you ask 100 different bird owners about the diets of their birds, you are likely to get 100 different answers. Since the advent of pelleted diets, there has been a debate over which is better, seed or pellets. The answer is neither. The objective for feeding birds is variety and this includes both seed and pellets along with whatever other healthy items you can get your bird to eat.

The question is often asked; how do you get a bird to eat a greater variety of fresh fruits and veggies? To understand the answer, it is good to ponder the following; if you offer a child an unlimited supply of pizza, chocolate cake, and broccoli, how much broccoli will be eaten? Birds, like children, will eat their favourite items first.

These are not always the most nutritious items and they tend to be the high-fat items. It is up to us to provide different foods in the proper proportions to encourage a broad healthy diet.

Feed Formulation

Feed formulation is the process of quantifying the amounts of feed ingredients that need to be put together, to form a single uniform mixture (diet) for poultry that supplies all of their nutrient requirements. It is one of the central operations of the poultry industry, in view of its role in ensuring good nutrition. Feed costs account for more than 70% of the total production costs for most types of poultry, so it is important that returns are maximised through use of adequate diets.

Feed formulation is a central operation in poultry production, ensuring that feed ingredients are economically used for optimum growth of chickens. It requires a good knowledge of poultry and feed ingredients. Most large-scale poultry farmers depend on commercial feed mills for their feeds, to obviate the need to do their own formulations or feed preparation. It is therefore essential that formulations are accurate, to ensure a large number of flocks are not adversely affected.

Typical Formulation

Feed formulation is both a science and an art, requiring knowledge of feed and poultry, and some patience and innovation when using formulae. Typical formulations indicate the amounts of each ingredient that should be included in the diet, and then provide the concentration of nutrients (composition) in the diet. The nutrient composition of the diet will indicate the adequacy of the diet for the particular class of poultry for which it is prepared (e.g. egg layers, meat chickens or breeders). It is common to show the energy and protein contents of the diet but comprehensive information on concentrations of mineral elements and amino acids may also be provided. Since lysine is a key amino acid for poultry, the concentrations of the other amino acids may be related to it. Because some nutrients interfere with the utilisation of other nutrients, relationships between nutrients, particularly amino acids, also show if each of the nutrients will be used efficiently.

Why are Diets Formulated for Poultry?

One of the reasons for formulating diets for poultry has already been identified, i.e. to produce a single uniform feed that can be delivered efficiently to poultry. The production rates expected on modern poultry farms also dictate that the dietary requirements of poultry be carefully identified and precisely met. Diet formulation enables the poultry industry to maintain some uniformity in levels of production. Industry nutritionists tend to use the same formulations over relatively long periods of time, so that the quality of the product (meat or eggs) remains stable over time.

The product quality can also be easily predicted if the same diet formula is used and all other factors remain unchanged. It is worth stating that chickens are able to select from different ingredients, placed separately, in order to grow and lay normally. Such a practice is known as choice feeding, and is practised on a limited scale by some poultry producers.

However, choice feeding is cumbersome and cannot be economically applied in large-scale modern commercial situations without modifications to the infrastructure.

How are Diets Formulated?

Although it does not appear to be a part of the feed formulation process, the first step in feed formulation is evaluation of ingredients. The requirements of the chickens for nutrients such as protein, mineral elements (e.g. calcium), and energy need to be identified. These requirements vary with age, type of chickens (layer or meat type), level of production, etc.

It is also important to obtain and compare the prices of feed ingredients, in order to reduce the overall cost of the diet. With this knowledge, mathematical formulae are used to derive the amounts of each ingredient that need to be included in the mixture, i.e. the diet. When using only a few ingredients, the formulae are simple, but a few ingredients are rarely able to supply all the nutrients that will meet the requirements of the bird, so several ingredients are used, requiring complex formulae. Some of these formulae have been built into computer programs, which enable the rapid processing of values that should be included in the formulation.

Computer programs also make it easy to check if nutrient requirements are met. It is important, however, to evaluate the diet in the laboratory or feed it to a small group of chickens to confirm the adequacy of the diet. This may not be necessary if the actual composition of the ingredients used is known and the actual nutrient composition can therefore be obtained. Although the average composition of many common ingredients is known, ingredients tend to vary between batches (e.g. in drought years cereals can be lower in quality).

Feed Ingredients

Feed ingredients for poultry diets are selected for the nutrients they can provide to poultry, the absence of antinutritional or toxic factors, their palatability or effect on voluntary feed intake and their cost.

Figure: *Wheat grains (Source: Aust. Chicken Meat Fed.)*

The key nutrients that need to be supplied by the dietary ingredients are energy, protein, vitamins and minerals. More information on measuring the nutrient composition of ingredients and the process of formulating poultry feeds is available in the section on feed formulation. In Australian poultry diets, energy is mainly provided by cereal grains, sometimes supplemented with relatively small amounts of fats. Protein is provided from both vegetable and animal sources, such as oilseed meals, legumes and abattoir and fish processing by-products. Some vitamins and minerals are provided from most ingredients but these are generally supplemented through a premix added to the diet. Diets may also contain additives for specific purposes. These are discussed in more detail in the section on feed additives.

The nutrient value of feed ingredients for poultry will vary depending on the species, variety or cultivar, the season, the source location, processing and storage conditions and the class of poultry being fed. Different classes of poultry differ in their ability to digest and absorb various nutrients, while the other factors will affect the intrinsic nutrient value of the ingredient.

Cereal Grains

The main cereal grains used in poultry diets in Australia are wheat, sorghum and barley. Other grains used to a lesser extent include rye, triticale and oats. Although the amounts and types of cereal grains included in poultry diets will depend largely on their current costs relative to their nutritive values, care must be taken to avoid making large changes to the cereal component of diets being fed to poultry as sudden changes can cause digestive upsets that may reduce productivity and predispose the birds to disease.

Figure: *Barley grains*

The quality of cereal grains will also depend on the seasonal and storage conditions. Poor growing or storage conditions can lead to grains with lower than expected energy content or contamination with toxin-producing organisms such as fungi and ergots

Nutrient Composition of Cereal Grains					
Ingredient	***Crude protein (%)***	***Metabolisable energy (kcal/kg)***	***Calcium (%)***	***Available phosphorous (%)***	***Lysine (%)***
Wheat	13.0	3153	0.05	0.20	0.5
Sorghum	9.0	3263	0.02	0.15	0.3
Barley	11.5	2795	0.10	0.20	0.4
Rye	12.5	2734	0.05	0.18	0.5
Triticale	15.4	3110	0.05	0.19	0.4
Oats	12.0	2756	0.10	0.20	0.4

Vegetable Protein Sources

The main vegetable protein sources used in Australian poultry diets are meal by-products resulting from commercial vegetable oil production, such as soybean, canola, cottonseed and sunflower meals and legumes, such as peas and lupins. Many oilseeds and legumes contain chemicals that can have detrimental effects when fed to poultry. These chemicals are called antinutritive factors. Some of these antinutritive factors can be destroyed by heat and so are not a problem in heat-treated meals. New cultivars of some oilseeds and legumes have been developed that are naturally low in antinutritive factors and so higher levels of the unprocessed grains can be included in poultry diets without ill-effect.

Nutrient Composition of Vegetable Protein Sources					
Ingredient	***Crude protein (%)***	***Metabolisable energy (kcal/kg)***	***Calcium (%)***	***Available phosphorous (%)***	***Lysine (%)***
Soybean meal	48.0	2557	0.20	0.37	3.2
Canola meal	37.5	2000	0.66	0.47	2.2
Cottonseed meal	41.0	2350	0.15	0.48	1.7
Sunflower meal	46.8	2205	0.30	0.50	1.6
Peas	23.5	2550	0.10	0.20	1.6
Lupins	34.5	3000	0.20	0.20	1.7

Animal Protein Sources

The main animal protein sources used in poultry diets are meat meal, meat and bone meal, fish meal, poultry by-product meal, blood meal and feather meal. Further information on animal protein sources in poultry diets is available in the section on animal protein meals.

Nutrient Composition of Animal Protein Sources					
Ingredient	***Crude protein (%)***	***Metabolisable energy (kcal/kg)***	***Calcium (%)***	***Available phosphorous (%)***	***Lysine (%)***
Meat meal	50.0	2500	8.00	4.00	3.6
Fish meal	60.0	2720	6.50	3.50	5.3
Poultry by-product meal	60.0	2950	3.50	2.10	3.4
Blood meal	80.0	2690	0.28	0.28	6.9
Feather meal	85.0	3016	0.20	0.75	1.7

Feed Additives

Modern intensive poultry production has achieved phenomenal gains in the efficient and economical production of high quality and safe chicken meat, eggs and poultry bioproducts. At the same time as making gains in production and efficiency, the industry has had to maximise the health and well-being of the birds and minimise the impact of the industry on the environment. The use of feed additives has been an important part of achieving this success.

Defining a Feed Additive

The diet of animals and humans contain a wide variety of additives. However, in poultry diets these additives are primarily included to improve the efficiency of the bird's growth and/or laying capacity, prevent disease and improve feed utilisation. Any additives used in feed must be approved for use and then used as directed with respect to inclusion levels and duration of feeding. They are also specific for the type and age of birds being fed. These guidelines are maintained by a government committee (Product Safety and Integrity; Australian Government Department of Agriculture, Fisheries and Forestry).

Common feed additives used in poultry diets include antimicrobials, antioxidants, emulsifiers, binders, pH control agents and enzymes. Sometimes diets will also contain other additives used in diets for humans and pets such as flavour enhancers, artificial and nutritive sweeteners, colours, lubricants, etc. Within each one of these classes of additives there can be dozens of specific additives manufactured and distributed by a wide variety of companies. Again, all ingredients and additives must be noted on the label and their use and inclusion levels meet the standards as defined by law. In some instances additives are added to the animal's diet in order to enhance their value for human consumption, but mostly this is accomplished by use of natural ingredients containing significantly higher levels of these nutrients that can be deposited directly into meat and eggs. This fact sheet will highlight a few important feed additives and their use in the poultry industry.

Growth Promotion Additives

Growth promoting hormones are not used in the poultry industry. The efficient growth and egg productivity of commercial poultry has been achieved over the last 50 years through traditional animal breeding techniques (genetic selection – not genetic engineering) and improved nutrition and management (including health and housing) practices.

Antimicrobials have been used extensively in intensive poultry operations to minimise disease and improve growth and feed utilisation. However, the industry is currently evaluating alternatives to chemical therapeutics. It should be pointed out that antimicrobial practices do not extend to production of commercial eggs (should a need for antimicrobials arise all eggs laid during the treatment and withdrawal period cannot be sold) and the meat industry must adhere to stringent guidelines with regard to drug withdrawal periods before marketing.

There is much controversy in regard to the impact of antimicrobials in animal diets on the development of resistant strains of microbes that could directly impact human health and carry over into meat and bioproducts as well as the negative impacts associated with their excretion into the environment. The European Union has moved towards a complete ban of in-feed antimicrobials for these reasons.

Development of alternatives to the present in-feed antimicrobials is an exciting area of current research world-wide. In all cases, it will be necessary to minimise disease challenges, strengthen the bird's natural defences (immune response, gut barrier/health) and optimise the diet to provide a balance of required nutrients for the bird's changing needs. All of these may be influenced by using feed additives.

Feed Enzymes

Enzymes are proteins that facilitate specific chemical reactions; following this the enzyme will disassociate and be available to assist in further reactions. Although animals and their associated gut microflora produce numerous enzymes, they are not necessarily able to produce sufficient quantities of specific enzymes or produce them at the right locations to facilitate absorption of all components in normal feedstuffs or to reduce anti-nutritional factors in feed that limit digestion.

Some cereal grains (rye, barley, wheat, sorghum) have soluble long chains of sugar units (referred to as soluble non-starch polysaccharides – NSP) that can entrap large amounts of water during digestion and form very viscous (thick gel-like) gut contents. Enzymes that are harvested from microbial fermentation and added to feeds can break these bonds between sugar units of NSP and significantly reduce the gut content viscosity. Lower viscosity results in improved digestion (more interaction of digestive enzymes with feeds and more complete digestion), absorption (better contact between digested feed nutrients and the absorptive surface of the gut) and health (reducing moisture in manure and nutrients available for harmful gut microflora to proliferate and challenge the birds (e.g. necrotic enteritis, a chronic intestinal disease caused by Clostridium perfringens, resulting in reduced performance, mortality and the main reason we currently use in-feed antimicrobials)).

Commercial enzymes are also produced that significantly reduce the negative effects of phytates. Phytates are plant storage sources of phosphorus that also bind other minerals, amino acids (proteins) and energy and reduce their availability to the bird. Ongoing research will develop enzymes that are more effective in maintaining function under a wider range of processing and digestive conditions. New enzymes may include those capable of reducing toxins produced during feed spoilage (mould growth in grains) and facilitating digestion of carbohydrates currently not available to simple-stomached animals (poultry, pigs, humans) such as cellulose, lignin and chitin. New feed additives are rapidly adopted by the poultry industry and have facilitated the development of significant new technology to advance the use and availability of in-feed enzymes.

Antioxidants

There are a variety of sources of reactive oxygen species (free radicals) in normal metabolism as well as those coming directly from feed ingredients. Oxidative stress can disrupt normal cellular function, damage tissues (also associated with the development of cancers) and reduce health status. Antioxidants bind these molecules and reduce their potential damage.

Acidifier

The Future

Many additives have been a normal part of diets for animals and humans. It is only recently that we have come to recognise and understand their importance in achieving high production and efficiency, maintaining health and wellbeing, improving product quality and safety and reducing the industry's impact on the environment. More work is required to further identify the positive effects of additives and minimise the negative effects they may have if not used correctly or if they interact with other additives or feed ingredients. In particular, additives will play an essential role in maintaining the health of poultry in an era of no pharmaceuticals.

3

Broiler Production Systems: The Ideal Stocking Density

The modern broiler house enables producers to have great control over the house environment. Birds can be placed at higher densities as long as the correct environment (temperature, ventilation, humidity) is provided. Factors to consider when determining stocking density include but are not limited to bird size, feeder space, drinker space, house dimensions, bird welfare, nutrition, breed, performance and economic return.

The ultimate goal is to maximise pounds of chicken produced per square foot while preventing production losses due to overcrowding. In many cases, producers have to settle for slightly reduced performance to achieve a satisfactory economic return. Another concern with increased stocking density is broiler welfare. Animal activist groups request that broilers be given more space during grow-out and cite behavioural and physiological stress as the reason.

Determining Stocking Density

Low Stocking density is calculated varies. Sometimes stocking density is reported using the number of birds per unit area or the amount of area per bird. For example broilers could be placed at .68, .70 or .75 square feet per bird. Currently many companies calculate stocking density by the pound. Instead of being expressed as the number of birds per unit area, density is calculated as bird weight per unit area. The benefit of using bird weight per unit area is that the standards are consistent and would stand true no matter how heavy the target weight. In short, once a company has determined how many pounds per square foot are needed to Optimise growth, development, feed conversion, livability, and economic return, they would decrease the number of birds per house as the target weight increases. Regardless of which method is used to report density, the same factors and issues are present.

Stocking Density Studies

The modern broiler does not appear to deal with stress in the same manner as those grown in the past. As a result, stocking density studies usually show that modern broilers perform better when given more space. However, studies are not always conclusive due to the numerous factors mentioned previously. Unfortunately, the broiler farmer cannot afford densities of 1 or 2 square feet per bird. Houses would not cash flow, growers would not achieve a satisfactory return, and birds would not be available to process.

Studies on stocking densities in broiler production have produced variable conclusions. Some studies show large benefits in reducing stocking density, while others show little or no differences. Bilgili and Hess (1995) conducted a study examining densities of 0.8, 0.9 or 1.0 square foot per bird. Body weight, feed conversion, mortality, carcass scratches and breast meat yield were significantly improved when birds were given more space.

Means within a column with different superscripts are significantly different (P<0.05). A study by Feddes et al. (2002) demonstrated that when bird density was reduced live body and carcass weights were decreased. However, bird uniformity was better at high densities. In that study, stocking density had no effect on mortality, breast meat yield, carcass grade, incidence of scratches, or carcass quality. It was concluded that high yield per unit area and good carcass quality could be achieved at the increased stocking density when adequate ventilation rates were provided.

Means within a column with different superscripts are significantly different (P<0.05). The impact that proper house environment has on broiler production at higher densities cannot be over emphasised. A recent study conducted by Dawkins et al. (2004) in the United Kingdom examined the effect of stocking densities on bird welfare in commercial facilities from 10 different companies. Stocking densities of 6.1, 7.0, 7.8, 8.6, 9.4 pounds per square ft (30, 34, 38, 42, and 46 kg per square meter) were compared. In addition to recording environmental conditions in the broiler house (temperature, relative humidity, ammonia, light intensity, and litter moisture), bird welfare was monitored through mortality, corticosteroid levels (a stress hormone), behaviour and health, with an emphasis on leg strength/structure and walking ability.

At higher stocking densities the birds grew slower, were jostled more and had reduced walking ability.

While stocking density significantly affected three of the measured variables, environmental management affected 17 of the 19 variables measured. It was concluded that while stocking density does affect broiler welfare, the management of the environment in the broiler houses were more important.

While stocking density certainly influences broiler performance and welfare, research indicates that housing environment is extremely important. Two of the studies discussed above demonstrate that it is possible to place broilers at higher densities, but that when this is done, broiler environmental management is crucial to optimising broiler performance and welfare.

Basic Introduction to Broiler Housing Environmental Control

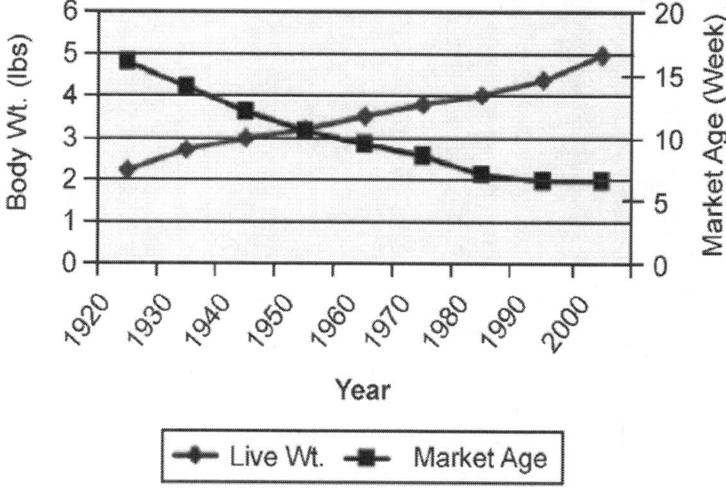

Figure: *Broiler Performance*

Genetics and nutritional improvements in broiler production have been extremely important to the efficiency of poultry meat production; however, the full genetic potential of broilers can not be reached unless the proper environment is maintained in the broiler house. The fast growing, modern broiler lines are more dependent on proper environmental conditions than birds from lines raised just a few years ago.

Construction

Broiler houses in the United States are constructed with wood or steel trusses and supports. The houses are clear span structures from side wall to side wall. The trusses are engineered to support the weight of the roof without the need of support posts that make it harder to catch birds and clean out the house. The floor is typically compacted dirt that is covered with bedding material (wood shavings, peanut hulls, rice hulls, sand, etc.). House dimensions are usually 40-50 ft wide, 400- 600 ft long with 8 ft high sidewalls.

Dropped Ceilings

To improve ventilation and reduce heating costs, most houses now have dropped ceilings. Dropped ceilings protect the trusses and ceiling insulation by acting as a vapour barrier. Dropped ceilings reduce the ceiling surface area and allows for the installation of ceiling insulation to reduce heat gain in during hot weather and heat loss during cold weather. Modern houses are well insulated with blown in cellulose or fibre glass batt insulation to reduce heat gain in the summer and heat loss in the winter. Insulation values of at least R-21 and R-7 are recommended in the ceiling and walls, respectively.

Solid Side Walls

Most houses are constructed with solid side walls rather than having open sides with curtains. This provides better insulation, reduces air leaks, provides better light control and allows the house to be heated more efficiently. The use of solid side walls provides a smooth surface compared to open sides walls with posts. This improves air speed during tunnel ventilation that will increase the cooling of birds next to the wall. Another trend in new construction is the building of larger houses. Houses as large as 70 x 600 ft have been constructed. If these houses prove to be cost effective, it is likely that most new houses will be constructed to larger dimensions in the future.

Heating

Maintaining proper temperature to promote efficient growth is key to profitable broiler production. Thus, heating a broiler house is extremely important from both a performance and economic standpoint. Chicks are not able to completely maintain their body temperature until approximately 14 days of age. During this time, it is crucial that floor temperature be maintained between 85-90 degrees F with minimum variation. The primary fuels used in heating U.S. broiler houses are propane or natural gas. Broiler heating systems include radiant brooders, pancake brooders, forced air furnaces and radiant tube heaters. Brooders and tube heaters project heat onto the floor. The hot air furnaces heat the air, which then heats the floor.

Hot air is lighter than cold air. This can result in stratification with the air being warmer at the ceiling than at the floor. Circulation fans are often used to move hot air from the ceiling down to the floor. Using circulation fans to mix the warm and cool air can result in as much as 30 percent fuel savings and may improve litter conditions as the warmer air on the floor helps dry litter. Paddle fans can also be used to mix air, but be careful to ensure that the chicks are not exposed to drafts.

Ventilation

Ventilation delivers fresh air and removes excess heat, moisture and undesirable gases from the broiler house. A typical ventilation system in a broiler house consists of fans, air inlets, evaporative cooling system and controller/thermostats. Houses are designed to deal with both cold and hot weather extremes.

Cold Weather Ventilation

During cold weather, negative pressure ventilation is used to provide fresh air, remove moisture and minimise heat loss. Fans exhaust air out of the house creating a slight negative pressure inside the house. Fresh air is pulled into the house due to the negative pressure and enters through planned air inlets that are installed either high on the house side wall or in the ceiling. These inlets are designed to direct air across the ceiling allowing it to mix with warmer air located there and to heat up before coming into contact with the birds. Newer houses use computer controllers to determine when the fans operate and for how long. The combination of controller and air inlets allows control of how much air enters the house and where it will enter and allows good air quality to be maintained while minimising heating costs.

Hot Weather Ventilation

During hot weather "tunnel ventilation" is used to keep birds cool. Tunnel ventilation systems consist of fans at one end of the broiler house and large air inlets at the opposite end. The fans pull air the length of the house at a velocity of 500 feet per minute. Tunnel ventilation removes heat from the building rapidly and creates a wind chill that provides additional cooling for the broilers. When tunnel ventilation alone is not sufficient to cool the broiler house, the evaporative cooling system is activated. Energy in the form of heat is used to evaporate water lowering the air temperature. Originally, evaporative cooling was accomplished using fogging systems located inside the house. The fogging nozzles provided a fine mist of water that evaporated, thus lowering the air temperature. Occasionally there were situations when this system was not used correctly. As a result the air sometimes became saturated and all of the water did not evaporate, which led to wet litter problems. This problem was corrected by moving the evaporative cooling system outside of the house. Fogging systems were placed on the end of the house where the air enters. The fogging nozzles sprayed a fine mist of water onto fluted/perforated pads. The air was drawn through the pads where water was evaporated and the air temperature was reduced. This system also water being wasted as it dripped off the pads.

Recirculating evaporative cooling systems have become popular as a solution to this problem and is the primary evaporative cooling system being installed currently. With this system, water runs through a perforated pipe at the top of the cool cell pads. Water runs down and through the pad soaking it. Any water that is not evaporated is caught in a trough at the bottom of the pad that delivers the unused water back to a reservoir to be pumped through the system again. Depending on environmental conditions (temperature, humidity), incoming air temperature can be lowered 10 degrees F or more.

Controlling House Environment

Almost all modern broiler houses rely upon electronic controllers. Through the use of controllers, it is possible to keep house temperatures within five degrees of the desired temperature regardless of outside temperature. This makes it possible to keep the birds comfortable so they are not diverting energy from growth to stay warm or cool. The controller monitors house environmental conditions and adjusts the heating, ventilation and cooling equipment as necessary to keep temperatures constant. Today, controllers can monitor temperature in six or more locations throughout the house. Humidity can also be monitored, although adjustments to heaters and fans are usually done on a temperature basis.

As the house temperature fluctuates, the controller will turn on the brooders or fans as needed. The controller operates equipment in the house including: brooders, fans, inlet machines, curtain machines, evaporative cooling systems and lights. Many controllers also allow house conditions to be monitored remotely. Using a computer and modem, a grower can call into the farm and check the temperature and humidity, as well as, which heaters and fans are operating in all houses. If needed, changes in the environmental settings can be made remotely using the computer.

Alarms and Generators

The importance of the maintaining a comfortable and stress-free environment for the birds cannot be overstated. Modern broiler housing can provide the environment needed to Optimise broiler performance, but this is entirely dependent on electricity and the proper operation of house equipment. It is difficult for a farm manager to be present 24 hours a day, every day that birds are in the house. Therefore, it is important to have an alarm system installed to let the farm manager know when something goes wrong in the house.

While the system will not correct the problem itself, its main purpose is the get someone into the house to evaluate and correct the problem. Alarms are used to notify if there is loss of power or if the house internal temperature gets too high or too low in relation to the desired temperature. The alarm system will activate a siren, usually located at the facility, to alert anyone close by and an automatic phone dialer and/or pager to notify the farm manager while he or she is away from the farm. In the case of power loss, emergency generators are used to operate ventilation, feeding and watering systems to prevent catastrophic losses. The emergency generator should have the capability of automatic power switch-over and be capable of delivering service for extended periods of time to operate the systems mentioned above.

Research on improving broiler housing is ongoing. Energy costs are becoming more significant to the grower's bottom line and housing construction, equipment and operation will be paramount in helping to make sure the houses are operated as efficiently as possible. As technology and equipment is redesigned and developed, researchers will continue to examine how broiler housing can be heated, cooled, and built in such a way that modern broilers continue to reach their genetic potential using the most economical and efficient methods.

Nutritional Influences on Hatching Eggs

The importance of nutrition in optimising broiler breeder performance is well understood. However, the ultimate aim for broiler breeders is to produce chicks with maximum potential for subsequent performance and liveability. Recent research has demonstrated the importance of the early nutritional status of the hatched chick in enhancing performance. It is therefore vital that, in formulating diets for the parent birds, attention is paid not only to maximising egg output and hatchability but also to ensuring that the eggs produced contain adequate amounts of the nutrients vital for supporting the early development of the hatching chick.

The micronutrient content of the egg is particularly influenced by the maternal diet and these nutrients must be supplied in amounts that fully meet the needs of the developing embryo and the hatching chick. It is helpful to consider the mechanisms by which these nutrients enter the egg, for this gives insights into the requirement for the nutrient and also limits the inclusion of the nutrient.

Most vitamins are incorporated into eggs by specific transport mechanisms involving binding proteins. This is essential, because circulating levels in the maternal blood are not high enough to allow the amounts needed in the egg to accumulate by simple diffusion. An example of this is the incorporation of riboflavin (Rb) into eggs under the action of riboflavin binding protein (RBP). RBP is a complex glycoprotein with a binding site for Rb and other phosphate rich sites that bind to receptors on the oocyte cell wall and enable RBP, whether or not it contains bound Rb, to be absorbed into the yolk. Thus, riboflavin is actively incorporated into the developing yolk. It is also concentrated into albumen by a mechanism involving transfer from the plasma to a RBP synthesised in the oviduct.

The concentration of RBP in hen plasma is constant and the fractional saturation of RBP with Rb increases with the dietary supply of Rb. Maximal saturation of RBP is about 80 per cent in the yolk but is lower in the albumen (about 40 per cent), perhaps because of a limitation in the plasma to oviduct transfer of Rb. However, once RBP is maximally saturated with Rb, further increases in the dietary supply are ineffective in increasing the egg content of Rb. Little, if any, free Rb is found in either yolk or albumen.

There are other binding proteins for other vitamins, though some of the principles may vary. Biotin is incorporated into yolk attached to two biotin binding proteins (BBP-I and BBP-II) and into albumen attached to a third, avidin. However, biotin can induce the production of BBP-II so that the total concentration of BBP increases with dietary biotin content, until a limit is reached to the production of BBP.

Incorporation of biotin tends to plateau at this point, though some additional free biotin can appear in yolk suggesting that some diffusion of biotin from plasma occurs at higher dietary biotin concentrations. Trace minerals are also incorporated into eggs by proteins, for example zinc is incorporated into yolk bound to vitellogenin.

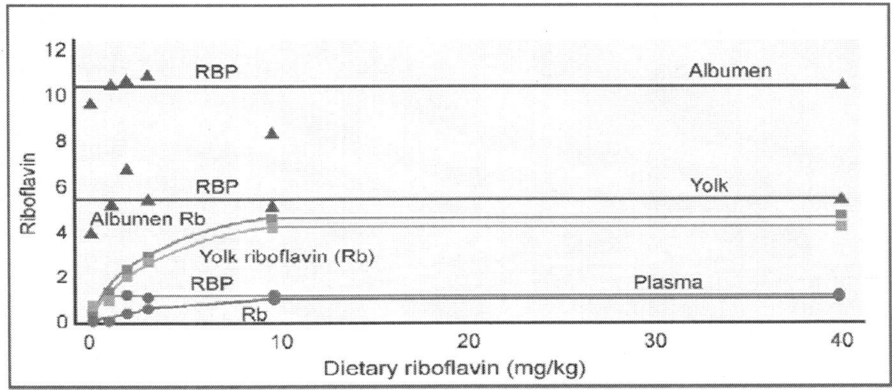

Figure: *The effect of dietary riboflavin on the RBP and riboflavin content of yolk, albumen and plasma*

The situation regarding the optimum incorporation of vitamin D and its metabolites into eggs is more complex. Vitamin D binding proteins can bind both cholecalciferol (vitamin D3) and 25-hydroxyvitamin D3 (25-D, available commercially as HyD) and transport these molecules into yolk. The chick embryo can produce the liver 25-hydroxylase and renal 1- and 24-hydroxylases necessary for the production of the more active metabolites, including 1,25- dihydroxyvitamin D (1,25-D). In contrast, 1,25-D itself is poorly incorporated into eggs and feeding only this form of vitamin D to hens does not sustain hatchability.

Further injections of 1,25-D into eggs only partially restores hatchability and there is evidence that another di-hydroxylated metabolite, 24,25-dihydroxyvitamin D, is needed in addition to 1,25-D for full hatchability. Thus, both cholecalciferol and 25-D in the diet of the hen can meet the full needs of the hatching chick for vitamin D. HyD has been assessed widely in breeders as an extra feed supplement in addition to normal cholecalciferol. Small scale experiments under laboratory conditions may not always give statistically significant responses but industrial experience in a number of countries is suggesting that commercially important responses can be achieved in egg production, shell quality or hatchability. This may suggest that the additional vitamin D status is beneficial under the more stressful conditions often experienced in large commercial flocks. The additional carryover to hatching chicks can also be helpful for their performance, not least in enhancement of early innate immune function. Carotenoids constitute a large group of lipid soluble compounds that are not traditionally thought of as nutrients. However, they have antioxidant and immunomodulatory functions and are important for pigmentation.

There is also a report that canthaxanthin supplementation (6mg Carophyll Red/kg) of a breeder diet based on white cereals improved hatchability. Carotenoids can be incorporated into eggs from the diet of breeders in a dose-related manner and are carried over into the hatching chick. If hens have a low rate of egg production, carotenoids are accumulated in tissues and skin and, indeed, the presence of highly pigmented skin on a hen can be used to identify poor layers within a flock.

The incorporation of carotenoids into egg yolk can involve transport in different lipoprotein fractions and perhaps also specific binding proteins but appears to have a synergistic effect on incorporation of other fat soluble compounds into the egg and also in the young chick. Thus Surai et al. (2003) have shown that addition of canthaxanthin (24mg/kg) to a breeder diet resulted in an increase in vitamin E uptake from the diet and transfer to the egg yolk. This was reflected in an increase in the liver vitamin E concentration of day-old chicks and a reduced rate of vitamin E depletion from liver over the first seven days post-hatch.

Canthaxanthin has been shown to react with peroxyl radicals with adduct formation, so the decreased rate of vitamin E depletion may have been attributable to canthaxanthin having a sparing effect on vitamin E by directly scavenging free radicals. Carotenoids can thus modulate antioxidant systems and decrease tissue susceptibility to oxidation. Other recent research has shown maternal effects of carotenoids on carotenoid uptake in growing chicks. Koutsos et al. (2003) compared carotenoid uptake in chicks hatching from eggs that were carotenoid replete or depleted through manipulation of the maternal diet. Chicks hatching from the replete eggs had higher tissue concentrations of carotenoids and showed much greater efficiencies of incorporation of carotenoids into immune tissues such as bursa and thymus and skin when fed diets containing lutein and canthaxanthin during the growing period. Provision of carotenoids in the maternal diet may thus benefit the immune system of the young chick. The maternal use of carotenoids can also improve the cost effectiveness of subsequent carotenoid use for pigmentation purposes by increasing the efficiency of incorporation of carotenoids into skin in growing broilers.

Understanding the Factors that Influence Broiler Breeder Flock Fertility

Flock fertility is dependent on the reproductive status of the birds (level of egg and semen production) combined with the bird's interest and capability of mating. Fertility usually increases from a low of 65-75% at the start of lay (23-24 weeks of age) and peaks at 95-98% at 35-37 weeks of age. Between 40-45 weeks of age fertility declines and the older the birds get the faster the decline in fertility.

What are the reasons for this decline? After 40 weeks of age, the breeder hen needs more frequent mating to sustain high fertility while at the same age, the rooster is less interested in mating. In situations where roosters are overweight, excessive breast fleshing may reduce their ability to complete matings. Overfeeding broiler breeder hens encourages additional fat pad deposition resulting in depressed fertility. High body weight of either sex increases bird mortality, causing a loss in egg production in the case of the hen or potential lower fertility by increasing the number of females per male. In addition, hen flocks that have poor egg production tend to have lower fertility levels throughout their productive lives.

How can We Sustain Fertility?

Of course, having good flock fertility is closely tied to proper feed restriction. As mentioned in the April 2001 *Hatchery-Breeder Tip*, weighing the birds on a weekly basis and adjusting the feed intake is essential to controlling body weight gains in these efficient converters of feed to meat. How the birds are weighed is important in how accurate the estimate of body weight. So, weigh weekly and develop a good method in order to get an accurate estimate of male and female body weight. While the estimate of body weight is being taken, handle the birds and judge the amount of breast fleshing and fat pad deposition. Roosters with a large amount of breast fleshing will have difficulty mounting and staying on the hen long enough to inseminate semen into the oviduct. Hens that are overweight tend to have large fat pads and are likely carrying large amounts of breast fleshing. Both of these physical attributes will reduce egg production and increase mortality (especially in hot weather), while obesity problem reduces the hens ability to store spermatozoa decreasing her fertility.

Ability to keep both sexes on target body weight is closely related to rooster exclusion from the hen feed trough. Each feeding system is slightly different and flock managers learn over time which houses have the tightest exclusion systems. One way to validate how much stealing is going on is to occasionally be present at feeding time. Using an exclusion grill with a horizontal bar across the top improves rooster exclusion particularly if the roosters have most or all of their comb (no dubbing). The horizontal bar also helps keep the grill in good shape, reducing places where roosters can steal feed. If the grill does not have a horizontal bar, PVC pipe placed in the top of the grill will have a similar effect. NozBonz have also helped many flock managers control rooster body weight; however, these flocks must be monitored closely which includes weekly weighing.

Monitor the sex ratio on the farm at bird placement in the hen house and about every month thereafter. Are you losing too many roosters and at any particular age? Are the roosters that remain on the farm in functional condition (good feet and legs, good body weight, or low number of injured males)? The right sex ratio is breed and condition dependent. The more aggressive the breed the fewer males needed. The better the body condition the more aggressive the male will be and again less roosters are needed. In general 7-9 roosters per 100 hens will usually net good fertility results. If your male to female placement ratio in the laying house is too low, mating frequency will likely be reduced. Additional roosters can be added they are available from other flocks. Having too high a placement ratio can cause the young hens to avoid the scratch area giving the roosters less opportunity to mate and reducing flock fertility, while increasing the incidence of male aggression (high rate of mortality and injury in both sexes).

After 40 weeks of age the addition of young roosters to the flock (spiking with 2%), is necessary to sustain high fertility if the flock will be held through 65-67 weeks of age. What should happen when the young males are added? If the older roosters are in good physical condition (not over fleshed or injured), mating frequency of the older rooster will increase for 4-5 weeks after spiking. The young males will have a high mating frequency but a low efficiency of mating the first 1-2 weeks after spiking. However, if the young males acclimate to the hen house (male feeder not too high, large enough body size to compete with older males, and mature enough at 26-28 weeks of age), their mating frequency will surpass the older roosters increasing the overall mating frequency and flock fertility. There are several ways to err with a spiking programme. The most critical mistakes are to ignore the health and parasite status of your spike males, use of males with poor physical condition, use of immature spike males, adding spike males to a high sex ratio situation, or starving the spike males by having the feeder too high.

Interspiking is a new method of improving fertility that has been successful for some integrated operations where young males are not available or biosecurity concerns will not allow the addition of roosters from another source. Interspiking is moving roosters from one house on a farm to another house on the same farm. This requires some labour to catch and release, but little else. Fertility will increase very quickly (as there is no acclimation period for these roosters), but fertility will decline in about 4-8 weeks as you are still dealing with an older male. Some producers will interspike several times during the life of the flock. If fertility is not consistently high from flock to flock, review your management practices and try something new.

Temperature Variation in on-Farm Hatching Egg Holding Units

Broiler breeder hatching eggs are commonly held in storage facilities at the breeder farm anywhere from one to four days and again at the hatchery until placed in the setters. In the poultry industry, some pre-incubation of hatching eggs following oviposition and during storage is inevitable, yet efforts should be made to reduce this occurrence. With the continued development of this industry there have been tremendous advances which have improved the available equipment to maintain hen house temperatures, and the quality of egg transportation vehicles and egg storage facilities in the hatchery. However, with this improved technology, on-farm egg storage facilities have been largely neglected which has made it extremely difficult for producers to maintain constant egg storage room temperatures at the farm level. While one purpose of egg storage is to accumulate eggs to meet the demand for chicks and to best utilise hatchery facilities, ultimately the goal is to arrest further embryonic development while maintaining embryo viability.

While an egg storage temperature of 68°F (20°C) is the most commonly practised industry recommendation, the actual on-farm egg storage temperature can range from a low of 60°F (15.6°C) up to 75°F (23.9°C). The range in egg storage temperature from one farm to the next is often due to different management programmes, while day-to-day fluctuation within the same company is a result of poor egg storage facilities that are unable to maintain a constant storage temperature. Hatchery egg storage conditions have been evaluated in the past, with recommendations presented to reduce losses in hatchability. However, research regarding egg storage at the breeder farm is limited and incomplete. The objective of this study was to determine the effects of oscillating and variable on-farm egg storage temperatures on hatchability and embryo viability in commercial broiler breeder flocks.

Egg Storage and Hatching Procedures

A total of 4320 hatching eggs were obtained from the University of Arkansas's Broiler Breeder Research facility and placed into two separate egg storage chambers, with all eggs stored at a control temperature of 70°F (21.1°C) for 0-24 hours. After the initial 24-hour storage period, eggs were divided into 864 egg lots and assigned to treatment groups. One group of eggs remained at 70°F for the entire 72-hour storage period (control). Four other groups were moved to separate storage chamber with temperatures set at either 66°F (18.9° C), 68°F (20.0°C), 72°F (22.2°C) or 74°F (23.3°C) to represent Treatments 1, 2, 3 and 4, respectively.

Eggs were stored at these temperatures for an additional 24 hours for a total of 48 hours of storage time. Then eggs stored at 66°F were stored at 74°F, eggs at 74°F were stored at 66°F, eggs at 68°F were stored at 72°F, and eggs at 72°F were stored at 68°F for an another 24 hours for a total storage time of 72 hours. After 72 hours of storage, all eggs were returned to 70°F.

This design ensured that all eggs in this experiment were held at an average of 70°F for the entire three-day "on-farm" egg storage time period. To summarise this design, all hatching eggs from the different temperature treatment groups were subjected to either a 2- or 4-degree F temperature fluctuation above and below the 70°F base temperature but were held at an average of 70°F. After the storage period, eggs were transported to their original commercial breeder farm where they were placed directly on a commercial hatching egg transportation truck and sent to a commercial hatchery for incubation. No treatment or special care took place after the on-farm storage period. The hatchability of eggs subjected to a 2°F temperature change from 70°F was reduced by nearly 2% compared to the control group (74.69 vs. 76.47% hatch, respectively). Eggs that underwent a 4°F temperature change had nearly a 1% loss in hatch as compared to the control group (75.61 vs. 76.47%, respectively). It is interesting to note that the greater temperature variation did not necessarily result in a greater loss in hatchability.

However, regardless of whether the temperature variation was 2 or 4°F, all hatching eggs used in the study moved from the hen house at about 80°F to the 70°F storage chamber for 24 hours. Eggs that then increased in temperature for 24 hours and decreased for another 24 hours before increasing again to 70°F (i.e. 70°F-Δ-∇-Δ) experienced a significant drop in hatchability as compared to the control (3.55% and 2.16% loss in hatch, respectively). Eggs in this group experienced multiple changes in temperature from the hen house to the hatchery. From the time of lay, these eggs decreased in temperature to 70°F then the temperature was raised for 24 hours, then lowered for 24 hours, then raised for 24 hours, then lowered as they were moved to the hatchery (67°F) then raised when moved to the setters (three periods of decreasing temperatures and three with increasing temperatures).

Eggs that were stored at 70°F then decreased in temperature for 24 hours, then increased after 48 hours then were returned back to 70°F (70-Ä-d-Ä) experienced no difference in hatchability and less than 1% loss in hatch of fertile. Eggs in this treatment group basically underwent one change in direction of the temperature they were subjected to from the time they were laid until the eggs reached the commercial hatchery. These eggs decreased in temperature after lay to 70°F, then the temperature was decreased again for 24 hours, then increased for 24 hours, then decreased for 24 hours, then decreased again as they were moved to the hatchery (67°F) then raised when moved to the setters (two periods where temperatures were decreasing and two with increasing temperatures). Each time the internal temperature of the egg is elevated to near 75°F, metabolic activity is again initiated and embryo development ensues only to be slowed again during additional egg cooling). While cooling hatching eggs is necessary, starting and stopping embryo development weakens the embryo and reduces its viability. The ideal situation is for hatching eggs to undergo only two temperature direction changes; one from the hen to the lowest temperature point at the commercial hatchery egg storage facility and the second temperature direction as eggs are moved into the egg setters.

It is well known that most hatchability problems are a result of poor fertility. However, when egg production is attained and the flock maintains high levels of fertility, how we care for hatching eggs can have a tremendous effect on overall hatchability. While current industry recommendations vary from 63°F to 70°F for on-farm egg storage, data from this research indicate that variations in on-farm egg storage temperatures of as little as 2°F can reduce hatchability by as much as 3.5%. Experience from evaluating current on-farm egg room temperature values indicates that variation in the actual temperature and the set temperatures are great and often exceed those parameters established in this study. Therefore, regardless of the equipment in the breeder house and the hatchery facilities, hatchability is routinely lost in commercial hatcheries due to neglect of the on-farm egg storage facilities.

Vitamin Levels for Modern Cost-Efficient Broiler Meat Production

Vitamins are micronutrients that participate in numerous organic metabolic processes and are therefore indispensable for excellent animal health and productive performance.

When compared to other nutrients, there are very few studies carried out in the last years to estimate the optimum levels of vitamins for broilers, and there is a huge variation in the levels used commercially. Most levels recomme-nded by the NRC (1994) were based on old studies, performed under controlled conditions and using the minimum levels to avoid signs of deficiency, not evaluating the best performance under the challenge conditions found in the field. Moreover, the modern breeds have higher growth and production rate, and have higher nutritional requirements to express their genetic potential. Besides production rates, other parameters are also presently evaluated to determine the vitamin requirements, as immunity, animal welfare, carcass characteristics, microbiological analysis, etc. Supplementation with higher levels than the minimum recommendations result in higher production performance, better health, welfare and carcass quality. In this sense, vitamins are micronutrients that take part in almost all organic metabolic processes and are vitally important for achieving good performance and health. Deficiency of one or more vitamins can lead to multiple metabolic disorders, resulting in decreased productivity, delayed growth, reproductive problems and/or decreased immunity.

Vitamins are divided into two groups based on their solubility in lipids (liposoluble) or water (hydrosoluble). The liposoluble include vitamins A, D, E and K, and the complex B vitamins (B1, B2, B6, B12, folic acid, nicotinic acid, pantothenic acid) and vitamin C are classified as hydrosoluble. In general, liposoluble vitamins have specific functions in the development and maintenance of tissue structures, while the hydrosoluble vitamins participate in catalytic functions or act as control mechanisms of the metabolism, as coenzymes (AWT, 2002).

The literature shows a large variation in vitamin levels used in commercial supplements for broilers (Rutz et al., 1999; Barroeta et al. 2002; Leeson, 2007). Therefore, there is a great interest in new studies to determine the levels that provide the best economic return, without interfering with the production performance of birds.

Vitamin Recommendations

When the nutritionist is considering the need for vitamin supplements, several factors have to be taken into consideration, as they can demand changes in the birds' requirements. These factors include breed, sex, management practices, development status of the bird, stress and diseases (Moreira, 2002). But there are also factors related to the feed as ingredients, energy level, processing, storage and vitamin sources. Studies are being performed on the use of higher levels of certain vitamins to improve the nutritional value and meat quality for the consumer.

The vitamin recommendations suggested by international research bodies as the National Research Council (NRC), Agriculture and Food Research Council (AFRC) and Institut National de Recherche Agronomique (INRA), and the Brazilian recommendations, as the Brazilian Tables for Poultry and Swine, are important foundations to estimate the levels that should be used in different production stages. However, they only present the minimum requirements, which are usually not sufficient under field conditions and have low correlation with the levels that presently used by the industry. Most studies to determine the vitamin requirement of broilers were performed under controlled experimental conditions using purified or semi-purified diets. These diets are highly digestible, their nutrients have high bioavailability, and include ingredients that not usual in feeding broilers, as isolated soy protein or casein (protein sources), and dextrose, starch or sucrose (energy sources), which also demonstrates the low correlation with the field situation (Leeson, 2007). Moreover, few trials have been performed in the last 30 years to estimate the vitamin requirement of broilers with higher genetic potential for growth, with more than 20% improvement in feed conversion (higher weight gain in a short period) (Perez-Vendrell et al., 2002) and a 87% increase in daily weight gain (from 26.8g/day in 1970 to 50g/day in 2000) (Barroeta et al., 2002).

Besides deficiency signs and/or weight gain and feed conversion, new parameters are being presently evaluated to determine the vitamin requirements of broilers, as immune response, animal welfare and quality of the final product (meat and eggs). The goal is to improve the aspect and nutritional value of the product and lengthen its shelf life. Higher vitamin levels have been used in the diet of broilers to compensate for variations in intake, bioavailability of the vitamins in the diet, antinutritional factors of the feedstuffs, stress (temperature, stocking density, management practices, diseases, etc). These are some of the other factors that can prevent the minimum requirements of the birds from being met.

Evaluating two levels of vitamin supplementation for broilers, Castaing et al. (2003) concluded that the highest level (approximately two times the standard dose used by the industry) resulted in higher weight at 38 days (1,919 g) than with the lower level (1,878 g). Moreover, the deposition of vitamin E in the carcass was also higher (5.4 mg/kg in the group supplemented with 20-25mg vitamin E and 12.5mg/kg in the group supplemented with 240mg). Perez-Vendrell et al. (2002) obtained similar results when they studied two supplementation levels, under controlled conditions (12.7 birds/m^2) or under stress (16.4 birds/m^2). The best outcomes (weight gain, feed intake, breast yield, vitamin deposition in the meat) were obtained with the higher supplementation level in both stocking densities.

In this context, the supplementation can be based on the Optimum Vitamin Nutrition (OVN) concept, which is presented by DSM (2006) as the most adequate level (optimum level) of both hydrosoluble and liposoluble vitamins. The objective is to Optimise the health status, welfare and productivity of the animals, assuring efficiency in the production of quality food. Using this concept it is possible to determine four vitamin supplementation levels related to the degree of animal response.

Animal Response / Vitamins in the Diet

1. Deficient: vitamin supplementation below the required level, and the animal is under risk of developing clinical signs of deficiency as a result.
2. Sub-optimum: vitamin supplementation in amounts that are appropriate to prevent deficiencies as long as the animals are under adequate health, environmental and physiological conditions (low challenge). However, when the animals undergo any type of stress, the level of vitamin supplementation is not sufficient to prevent reduced performance or reproduction rates.
3. Optimum: contributes for the maximum expression of the productive potential of modern breeds under field conditions.
4. Special applications: besides their contribution to the maximum performance of the animal, these vitamin supplementation levels improve certain attributes as the quality pf the final product (meat and eggs) and immunity.

As for most nutrients, the vitamin requirements for broilers have probably undergone very few modifications in the last 30-40 years, as the nutrient levels required for maintenance are practically fixed and the composition of muscles / tissues are resistant to changes. There has been, however, an increase in the vitamin requirements for specific metabolic processes, as the immune responses related to performance expectations of broilers under high stocking densities imposed by today's commercial rearing conditions. As an example, there has been a linear decrease in vitamin E intake in the last 20 years, of 0.8%/year/kg weight gain, considering 20 IU vitamin E/kg feed and a feed conversion of 1.0 in 1987 and 1.7 today (Leeson, 2007).

Change in the Parameters of Vitamin Levels Determination

Much emphasis is given presently to nutrients with nutraceutical functions, mainly the vitamins, as they play an important role in promoting health, well-being and immunity.

The classical dose-response curve evaluation, often used to estimate the requirements of other nutrients, does not seem to be the most adequate for vitamins. Older studies evaluated the minimum requirement of vitamins needed to prevent the animal from presenting deficiency signs or they evaluated basic performance variables as weight gain, feed conversion and mortality. As most of these trials were performed under controlled environmental conditions (low challenge), the levels that were obtained are very little representative in practice. To find the optimum levels for broilers reared under industrial conditions, other factors beside performance should be evaluated, as carcass characteristics, breast yield, microbiological quality and immune response.

Vitamin Levels used in Brazil

Table shows the relationship between average levels found in the vitamin supplements used by the major Brazilian broiler companies in 2008, the levels based on the Optimum Vitamin Nutrition (OVN) concept for broilers and also the levels used in Brazil in 2005 (Nascimento et al., 2005). It can be seen that there was an increase in the vitamin levels used by the Brazilian industry during the growth period when 2008 and 2005 figures are compared. It can be concluded that the companies are aware of the need for a higher vitamin supplementation to follow the birds genetic development, although the levels being used are still bellow those put forward by the broiler breeding companies and the OVN levels recommended by DSM (2006).

Cost of the Vitamin Supplementation

The physical and chemical characteristics of vitamins are important aspects and should be taken into consideration not only by the premix producer but also by the final consumer, the buyer of the vitamin premix. Both should demand that high quality raw materials are used in the production of the premix. The increase in the vitamin levels represents only a 0.5% increase in the total cost of the feed. However, if the quality of the vitamin premix is not assured, the supplementation of marginal vitamin levels can cause serious losses in the animals performance.

According to the authors, changes in the research parameters and methods are necessary to obtain more accurate results, applicable to practical conditions. As to the vitamin requirements of today's broiler breeds. They are related to optimum performance, final quality of the product and financial return. Besides the production rates within this context, there are also important variables to be analysed as immunology, interaction between vitamins and other nutrients, composition, meat quality and food safety, performance under stress conditions, effect of supplementation of the breeder's diet on chick quality. New trials involving the vitamin requirements of modern broiler breeds reared under industrial conditions are needed. These trials should include some form of stress to the birds in order to be more similar to the industry' real situation. They should also take into account variables such as vitamin levels in the final product (meat and eggs), shelf life and organoleptic characteristics.

Chicken Embryo Malpositions and Deformities

Detailed information describing the categorisation and incidence of embryo malpositions and deformities in commercial poultry is not readily available. Additionally, there is often little consistency in these data among hatcheries. Any decrease in the number of usable chicks may result in substantial economic loss to poultry integrations. In a typical hatch, it is common to lose about 1-2% of the chicks due to deformities and malpositions.

Deformities manifest during the process of embryo development, while malpositions occur in the last week of incubation before hatch. At a commercial hatchery over a 5 year period, more than one-half million eggs had been broken out for quality control purposes and many thousands of unhatched embryos had been examined to determine the frequency of the various deformities and malpositions.

The objective of this study was to determine the relative incidences of malpositions and deformities, and their economic impacts. Major factors affecting their occurrence will be explained. Obviously, in any population it is anticipated to encounter malpositions and deformities during embryonic development. However, the incidence must be within accepted limits and changes must be made when excessive losses occur.

Malpositions

Investigation has demonstrated that the incidence of embryos unable to hatch due to malpostions varies from 1.2 to 1.8%, with an average of 1.5%. Malpositioned embryos are unable to pip the eggshell and escape due to improper positioning within the egg in the hatcher. It is interesting to note that numerous malpositions have been described, with some embryos exhibiting only one form of malposition and others experiencing combinations of malpositions.

The majority of eggs with malpositioned embryos, as found in hatch residue, included embryos dead in shell, probably resulting from exhaustion and/or lack of oxygen. A smaller number of eggs contained live embryos trying to pip. Loss of embryos due to malpositions may be costly; therefore, it is important to routinely monitor the percent of the embryos not hatching. If the incidence due to malpositions exceeds the standard, corrective measures must be taken.

An embryo provided an optimum environment for development will position itself around 17-18 days of incubation for hatch. The proper position is with the *head under the right wing* with the head directed towards the aircell in the large end of the egg. The results of this study demonstrate that malposition # 6, which is beak above the right wing, constitutes almost 50% of the malpositions, followed by position # 5, feet over head with a frequency of 20%.

There are numerous reasons that malpositions occur. In a *normal* population, the incidence should not exceed 2.0%. If the incidence is elevated, breeder and egg management practices must be investigated and appropriate changes made to resolve the problem. Common reasons for increased incidences of malpositions are:

- Eggs are set with small end up. As part of a monitoring programme, check eggs in the egg room or in the setters to ensure that eggs have been set correctly.
- Advancing breeder hen age and shell quality problems.
- Egg turning frequency and angle are not adequate. Proper frequency of turning through a 45 degree angle assists the embryo to position for hatch. The standard turning rate in the setter is 1 per hour.
- Inadequate percent humidity loss of eggs in the setter. Acceptable weight loss of eggs from setting to transfer is 11-14%.
- Inadequate air cell development, improper temperature and humidity regulation, and insufficient ventilation in the incubator or hatcher.
- Imbalanced feeds, elevated levels of mycotoxins, and vitamin and mineral deficiencies.
- Exposure to lower than recommended temperatures in the last stage of incubation.

Table: *Incidence of the common malpositions*

Malposition #	Description of the malposition	%
1	Head between thighs	12.5
2	Head in the small end of egg	7.5
3	Head under left wing	7.5
4	Head not directed toward air cell	4.5
5	Feet over head	20.0
6	Beak above right wing	48.0

Deformities

In any animal population during embryonic development, there is a *predictable* incidence of embryos that die or are not able to hatch due to deformities. Based on this comprehensive investigation, data demonstrated that the percent of deformed embryos ranged from 0.22 to 0.30% of the total hatch. These findings suggest that hatchability declines on the average of 0.25% due to malformed chicks. A combination of deformities and malpositions can be manifested simultaneously. The incidence of common deformities observed from embryos at 15 to 21 days of incubation. The most common deformities are those of exposed brain (29%), without eye(s) (25%) and with beak abnormalities (+/-35%).

Table: *Incidence of common deformities*

1	Exposed Brain	29
2	Without eye(s)	25
3	4 legs	10
4	Deformed beak	27
5	No upper beak	8
6	Deformed twisted leg	1

The occurrence of deformities in a population is considered acceptable as long as it does not exceed the 0.30% limit in an average normal hatch of 85%. As with malpositions, there are many factors that contribute to an increased incidence of deformities including:

- Male and female age, cross and breed. Younger breeders and use of fresh sperm reduce incidence of deformities.
- Egg management and storage practices. Use caution to prevent physical abuse of fertile eggs. Set eggs soon after lay, not exceeding 3-4 days of storage.
- Ambient factors, especially temperature and humidity, that affect embryo development. An elevated incubator temperature accelerates embriogenesis, and organs might not grow in synchrony.

  High machine temperatures are associated with brain and eye development problems, while lower than normal temperatures retard growth.
- Breeder diets deficient in macro-nutrients such as proteins, or micro-nutrients such as vitamins and minerals. An embryo grows using the internal egg nutrient content, including the yolk, shell and albumen of the egg. Hens fed vitamin-deficient diets may produce embryos and chicks that exhibit classical nutritional deformities and an increased percent of malpositions. Dietary deficiencies may result in sudden declines in hatchability.

The objective is to produce the maximum number of healthy chicks from eggs set. The percent hatchability in the commercial poultry industry ranges from 78-88%. Many variables affect the level of success, including environmental temperature and humidity, lighting, body weight management, strain of breeder, etc. Normally, loss of about 1.8% of total hatch due to malpositions and deformities can be anticipated. However, if this is elevated, necessary corrective measures must be taken. The importance of a routine embryo diagnosis programme can not be overstated. Without such a programme and access to data generated, it is difficult to detect when increases in incidences of 0.5-1.0% deformities and malpositions occur. Thus, it is not possible to "know where to look" for the problem and make the necessary changes. The competent hatchery manager will consistently be able to obtain superior hatches by being able to identify even small problems and promptly resolve them. In most cases in the hatchery, problems with hatchabiltity are due to a combination of several unresolved smaller problems.

Multi-Flock Comparison of Broiler Feed Ticket Weights and on-Farm Feed Weights

Contract broiler producers do not manufacture or deliver feed to their farms nor do they purchase it outright. Because they do not directly pay for their feed, questions may arise concerning how much feed was delivered and if that feed was accurately weighed, even though contracts have provisions for growers to be present when feed and birds are weighed. Accurate feed weights are critical because under most broiler growing contracts, a major portion of producer pay is based on how much feed birds consume and how well birds convert that feed to meat. The ABRF recently harvested the 100th flock of broilers grown at the farm. The farm is capable of weighing the feed birds consume on a daily basis and thus was able to compare feed-ticket weights with on-farm feed weights for 100 flocks of birds.

Comparisons

The on-farm feed weighing system and procedures for daily feed weighing were described in detail previously by Tabler (2001). The system allows the ABRF the capability to weigh feed consumed at each of four broiler houses on a daily basis. After harvest, feed intake for the four houses are combined and compared with feed ticket weights received during the flock and feed weight charged to the farm on the settlement sheet.

Data indicate that on-farm weights and feed-ticket delivery weights never exactly matched but were for the most part, similar. For 81 of 100 flocks, weight differences favoured the producer.

Table: *On-farm bin scale weights versus feed ticket weights (flocks 1-50)*

Flock No.	*Flock dates*	*On-farm Feed Wts (lbs)*	*Scale Ticket Wts (lbs)*	*Difference (lbs)*	*Difference (%)*
1	11/19/90-1/14/91	853330	846900	6430	0.754
2	2/1/91-3/29/91	819520	814480	5040	0.615
3	4/15/91-6/9/91	814290	806240	8050	0.989
4	6/20/91-8/18/91	840360	886960	lightening damage	
5	8/29/91-10/23/91	865658	859360	6298	0.728
6	11/12/91-1/7/92	911938	903720	8218	0.9
7	1/23/92-3/16/92	802864	793960	8904	1.109
8	4/2/92-5/21/92	688720	683580	5140	0.746
9	6/8/92-7/30/92	757580	751230	6350	0.838
10	8/7/92-10/1/92	885928	881620	4308	0.486
11	10/15/92-12/10/92	967180	962810	4370	0.452
12	12/21/92-2/17/93	970436	962900	7536	0.777
13	3/2/93-4/29/93	973240	965190	8050	0.827

Contd...

Flock No.	*Flock dates*	*On-farm Feed Wts (lbs)*	*Scale Ticket Wts (lbs)*	*Difference (lbs)*	*Difference (%)*
14	5/11/93-7/6/93	875352	868970	6382	0.729
15	7/9/93-9/2/93	857972	853220	4752	0.554
16	9/17/93-11/11/93	984974	978570	6404	0.65
17	11/29/93-1/25/94	1072612	1062440	10172	0.948
18	2/10/94-4/6/94	948546	935060	13486	1.422
19	4/19/94-5/31/94	660784	655240	5544	0.839
20	6/9/94-8/3/94	748054	748560	506	0.068
21	8/5/94-9/14/94	588722	586160	2562	0.345
22	9/20/94-11/3/94	666354	664020	2334	0.35
23	11/15/94-12/28/94	671776	665860	5916	0.88
24	1/10/95-2/23/95	692770	686280	6490	0.937
25	3/7/95-4/19/95	578528	582980	4452	0.764
26	5/5/95-6/15/95	649266	644900	4366	0.672
27	6/29/95-8/9/95	618756	610200	8556	1.383
28	8/18/95-9/28/95	647574	641960	5614	0.867
29	10/13/95-11/22/95	613104	605720	7384	1.204
30	12/7/95-1/22/96	665134	671360	6226	0.927
31	1/26/96-3/7/96	557626	552940	4686	0.841
32	3/15/96-4/26/96	601490	595900	5590	0.829
33	5/9/96-6/20/96	598276	593240	5036	0.842
34	7/4/96-8/16/96	618418	606780	11638	1.882
35	10/31/96-12/10/96	685446	689340	3896	0.565
36	12/30/96-2/6/97	591834	581120	10714	1.81
37	2/24/97-4/7/97	663096	654200	8896	1.342
38	4/24/97-6/6/97	661088	652410	8678	1.313
39	6/26/97-8/18/97	858594	850380	8214	0.957
40	9/1/97-10/22-97	776572	770300	6272	0.808
41	11/7/97-12/30/97	839070	830120	8950	1.067
42	1/27/98-3/20/98	848298	843280	5018	0.592
43	4/6/98-5/27/98	777952	767860	10092	1.297
44	6/12/98-8/6/98	816662	813440	3222	0.395
45	8/18/98-10/12/98	866424	863020	3404	0.393
46	10/30/98-12/15/98	746540	695350	51190	6.86
47	1/8/99-3/1/99	818744	810900	7844	0.96
48	3/22/99-5/14/99	831298	820820	10478	1.26
49	5/31/99-7/27/99	933730	928680	5050	0.54
50	8/5/99-9/29/99	911550	901080	10470	1.15

Contd...

Flock No.	***Flock dates***	***On-farm Feed Wts (lbs)***	***Scale Ticket Wts (lbs)***	***Difference (lbs)***	***Difference (%)***
51	10/12/99-12/3/99	851880	856600	4720	0.55
52	12/20/99-2/8/00	784042	778900	5142	0.66
53	3/13/00-5/4/00	854550	845030	9522	1.11
54	5/15/00-7/11/00	930726	930940	214	0.02
55	7/21/00-9/12/00	853534	842980	10554	1.24
56	9/22/00-11/13/00	844766	841120	3646	0.43
57	11/28/00-1/19/01	784058	781980	2078	0.27
58	1/30/01-3/23/01	927512	916700	10812	1.18
59	3/29/01-5/10/01	660764	653700	7064	1.08
60	5/18/01-6/30/01	671108	659980	11128	1.69
61	7/5/01-8/17/01	727610	728360	750	0.10
62	8/30/01-10/10/01	681540	651560	29980	4.60
63	10/30/01-12/7/01	611030	608200	2830	0.47
64	12/21/01-2/6/02	903546	898850	4696	0.52
65	2/15/02-4/1/02	868838	866780	2058	0.24
66	4/11/02-5/28/02	930624	935990	5366	0.57
67	6/4/02-7/19/02	843580	831660	11920	1.43
68	8/5/02-9/18/02	770174	767240	2934	0.38
69	11/4/02-12/17/02	697376	697780	404	0.06
70	1/3/03-2/14/03	650214	649670	544	0.08
71	2/27/03-4/10/03	610242	608270	1972	0.32
72	4/29/03-6/10/03	612478	606510	5968	0.98
73	6/19/03-7/31/03	603640	603070	570	0.09
74	8/18/03-9/29/03	591556	589250	2306	0.39
75	10/7/03-11/18/03	685668	677240	8428	1.24
76	12/30/03-2/10/04	749558	752090	2532	0.34
77	2/23/04-4/2/04	610150	606040	4110	0.68
78	4/15/04-5/26/04	563054	561720	1334	0.24
79	6/3/04-7/17/04	645268	637870	7398	1.16
80	8/22/04-10/11/04	740508	733550	6958	0.95
81	10/17/04-11/29/04	713678	699580	14098	2.02
82	1/3/05-2/14/05	809018	809790	772	0.10
83	2/28/05-4/11/05	772700	766430	6270	0.82
84	4/25/05-6/3/05	647250	642160	5090	0.79

Contd...

Flock No.	*Flock dates*	*On-farm Feed Wts (lbs)*	*Scale Ticket Wts (lbs)*	*Difference (lbs)*	*Difference (%)*
85	6/13/05-7/22/05	617892	614490	3402	0.55
86	8/8/05-9/16/05	588314	587940	374	0.06
87	4/11/06-5/19/06	619420	619640	220	0.04
88	6/5/06-7/13/06	703208	704860	1652	0.23
89	8/1/06-9/21/06	1001448	1005120	3672	0.37
90	10/6/06-11/24/06	1002777	992780	9997	1.01
91	12/21/06-2/7/07	902726	912880	10154	1.11
92	2/26/07-4/20/07	947448	947020	428	0.05
93	5/15/07-7/10/07	1035918	1037400	1482	0.14
94	7/27/07-9/24/07	1085270	1088770	3500	0.32
95	10/8/07-12/3/07	1114396	1123690	9294	0.83
96	12/14/07-2/6/08	1003356	998160	5196	0.52
97	2/21/08-4/11/08	947624	945970	1654	0.17
98	4/25/08-6/13/08	906596	915510	8914	0.97
99	6/26/08-8/14/08	929254	942698	13444	1.43
100	8/22/08-10/10/08	950312	955475	5163	0.54
	Totals	***39558199***	***39429993***	—	—
	Average	***791164***	***788600***	***5454***	***0.70***

[1]Bold numbers indicate flocks when scale ticket weights were greater than on-farm feed weights.

For the first 50 flocks, the difference between on-farm feed weights and feed-ticket weights favoured the producer in 47 flocks by an average 1.02 per cent. The remaining three flocks favoured the integrator by an average of 0.59 per cent. The overall combined difference between the two systems on the first 50 flocks was 0.99 per cent in favour of the producer.

In the second 50 flocks, (51 through 100) the difference between on-farm feed weights and feed-ticket weights favoured the producer in 34 flocks by an average of 0.81 per cent. The remaining 16 flocks favoured the integrator by an average of 0.48 per cent. The overall combined difference between the two systems on the second 50 flocks was 0.70 per cent in favour of the producer, slightly lower than the first 50 flocks. Ten of the 16 flocks that favoured the integrator have occurred since 2006, after the ABRF was renovated. Several older load-cells at bins that weigh feed on-farm had to be replaced during that time. Many of these were original load-cells installed in 1990. This change in load-cells may partially explain the increased number of flocks that recently favour the integrator. Overall, differences between on-farm feed weights and feed-delivery ticket weights averaged 0.85 per cent for 100 flocks of broilers that consumed over 75 million pounds of feed.

Everyone's Best Interest

Every producer will eventually face the issue of feed being delivered a day too early and the bins not holding it all. By law, any feed that is returned by or picked up from a poultry producer must be weighed if feed weight is a factor in determining payment; the integrator must document and account for any returned or picked-up feed.

But sometimes mistakes occur. Growers are urged to keep track of feed tickets and know when something is out of the ordinary – for example, feed is delivered too frequently or not frequently enough. It is in the best interests of both producer and integrator to resolve any issues as they arise.

Waiting until the flock sells to try to resolve a questionable feed ticket from weeks ago may result in conflict. Service technicians should be contacted at the first sign of a potential problem. If feed was received but no ticket can be found, ask the service technician to provide a copy. Feed costs are nearly two-thirds of all broiler production costs.

These costs are important to growers because feed conversion largely determines how well each grower settles at the end of the flock. Integrators have millions of dollars invested in feed mills and feed processing equipment, feed trucks, labour costs and feed ingredients. This is, in part, why service technicians are always asking growers to manage feeder height correctly and to avoid feed waste when chicks are small and begin scratching feed out of the feed trays.

Safeguards

The USDA Packers and Stockyards Act (P&S Act) of 1921 is designed to promote fair competition and ensure fair trade practices in the livestock and poultry industries. This Act also protects contract poultry producers. P&S Act requires all scales used by integrators to weigh feed for purposes of payment and settlement be installed, maintained and operated to ensure accurate weights. The Packers and Stockyards Programme (P&SP) enforces the Packers and Stockyards Act (P&S Act). P&SP promotes accurate weighing in the live poultry industry in the following ways (USDA, 2008):

1. All scales used for weighing feed for purchase, sale, acquisition, payment or settlement must be installed and maintained in accordance to National Institute of Standards and Technology (NIST) Handbook 44 as incorporated by reference into the regulations.
2. All scales used to weigh feed for purchase, sale, acquisition, payment or settlement under the Packers and Stockyards Act (P&S Act) must be tested for accuracy by a competent agency at least each six months, and the reports of these tests forwarded to P&SP.
3. Any scale found to be inaccurate according to accepted tolerances, must not be used until it is repaired, re-tested and found accurate again.
4. Whenever the weight of feed is a factor in determining payment or settlement to a poultry grower when live poultry is produced under a growing arrangement, live poultry dealers must base payment or settlement on the actual weight of feed shown on the scale ticket. If the actual weight used is not obtained on the date and at the place of transfer of possession, this information must be disclosed with the date and location of the weighing on the accountings, bills or statements issued. If there are any adjustments to the actual weight, this information and the reason must be disclosed on the accountings, bills or statements issued.
5. Integrators must employ qualified scale operators. Integrators must require scale operators to comply with federal regulations for weighing feed for payment purposes.
6. Every live poultry dealer must keep all accounts, records and memorandum necessary to fully and correctly disclose all transactions involved in the business transaction, including the true ownership. The scale ticket is a legal document. Every record that is issued where weight is a factor of settlement depends.

Scales used to weigh feed must be attached to a printer, which should print weight values on a feed ticket. Producers should never receive scale tickets written by hand. In addition to safeguards at the feed mill, the P&S Act also requires that each scale ticket for feed, where weight of feed is a factor in determining settlement to a producer, must show (USDA, 2008):

1. Name of the company performing the weighing service
2. Name and address of the producer receiving the feed
3. Name, initials, or number of the feed weigher, or if required by State law, signature of the feed weigher
4. Location of the scale
5. Gross, tare, and net weight of each lot assigned to an individual producer
6. Date and time gross and tare weights were determined, if applicable
7. Whether the driver was on or off the truck at the time of weighing, and
8. Licence number of the truck or other identification numbers on the truck and trailer, if weighed together, or trailer if only the trailer is weighed.

Even though integrators are required by law to maintain accurate records, it is important for growers to retain feed tickets and maintain accurate records.

Alert the integrator at the first indication there may be a potential concern. The longer growers wait to report a problem, the harder it will be to resolve that problem. In most cases, problems can be quickly resolved when both producer and integrator have accurate records and act in a timely manner. In the unlikely event that a producer cannot satisfactorily resolve a feed issue, the P&S Act authorises the Grain Inspection, Packers and Stockyards Administration (GIPSA) to investigate complaints of possible violations. Producers may report possible violations of the P&S Act to GIPSA toll free at 1-800-998-3447.

Egg Shell Mottling and Hatchability

Hatchability Problems

Hatching egg quality parameters have become increasingly important as commercial broiler breeder producers attempt to maximise hatchability. The egg pack can be easily monitored and growers held responsible for sending too many poor quality eggs to the hatchery. However, even good quality eggs can be mishandled. When care is not taken the incidence of otherwise good hatching eggs sent to the hatchery in the form of upside-down, or filth covered eggs, which may cause contamination, or even slab sided eggs, will also reduce hatchability. When troubleshooting hatchability problems, traditionally producers have placed the blame in one of three areas, fertility, hatchery (incubation) conditions, or egg handling.

Obviously most of the attention is usually turned to the males in the breeder house and overall flock fertility. This is normal considering that the majority of actual hatchability related problems are directly related to poor fertility. Additionally, poor fertility is correlated with increased early embryo mortality which results reduced hatchability.

A second area often responsible for poor hatchability can be directly linked to actual hatchery or incubation conditions. Even with the modern technology available today, hatchery equipment can, and does, wear out and malfunction over time. Equipment maintenance is often more than a full time job when trying to manage a hatchery for optimum production.

A third area often responsible for reductions in hatchability is egg handling conditions and procedures. While it is obvious that we have much to learn in this area and that our 'tried and true' methods for egg handling may not be the best, that will be the focus of future articles. The purpose of this article is to address another area that is sometimes blamed for poor hatchability, namely egg shell quality.

Link between Broiler Intensification and Foodborne Pathogens Explored

Animal welfare could be a victim of the drive towards improving global food security, according to Professor Tom Humphrey of the National Centre for Zoonosis Research at the University of Liverpool. He was speaking on 'Endocrines and host-pathogen interactions' at a combined meeting of the British Society of Animal Science (BSAS) and the UK branch of the World's Poultry Science Association (WPSA) in Nottingham in April 2011.

Professor Humphrey continued by asking how much animal welfare matters when price is the driver of most poultry meat sales in the UK and elsewhere, and intensive production is the main reason why poultry meat can be sold cheaply. "If you change anything about an animal, you also change its associated micro-organisms, like Campylobacter," he warned.

The important question, he continued, is whether an unhappy animal poses a greater threat to food safety. According to Professor Humphrey, the changing pattern of human Campylobacter infections over the last 20 years – shifting from affecting mainly the young to older people over this period – is at least partially related to the industrialisation of chicken production. He continued that Campylobacter poses broadly two health threats. The first is high levels of surface contamination, with levels of up to the 10^9 being recorded. However, Professor Humphrey is more concerned about the second, the contamination of liver and muscle cells after the bacteria leave the gut. He said that 25 per cent of muscle samples have been found to contain Campylobacter and 75 per cent of chicken livers. This may be the result of contamination during the slaughter process, especially during evisceration and scalding, he said, but it is also possible the organism is changing on the farm, becoming pathogenic.

Most Campylobacter become invasive only when the chicken's health or welfare are compromised, for example, when the bird is stressed during catching or immuno-suppressed as the result of an endemic disease such as avian pathogenic *Escherichia coli* (APEC). However, some strains are always invasive, and leave the gut for the liver or muscle, causing lesions, he explained. It is Professor Humphrey's contention that Campylobacter, a common zoonotic pathogen worldwide, is becoming increasingly hard for chickens to control, and that it is the state of the host that influences both symptoms and outcome. "Animals are victims of their environment," he said, adding that the solution the controlling Campylobacter lies in keeping the animal's gut healthy.

Roles of Stress and Other Pathogens on Campylobacter Risk

Sources of stress in poultry include major physiological events such as egg-laying or hatching, mixing social groups, production-related conditions (including hock burns and breast blisters, thinning and transportation) and poor diets that inflame the gut mucosa. Stress will also cause inflammation of the gut and modulate immune function in ways that may benefit Campylobacter. Professor Humphrey suggested that the behaviour of Campylobacter in vivo is strongly linked to the health and welfare of the host.

This is based on the observation that noradrenaline has been shown to increase the growth of Campylobacter. Noradrenaline is released into the gut when an animal is stressed and it leads to changes in the gut that increase the availability of iron. It is known that Campylobacter competes poorly for iron with other gut microflora. Furthermore, *Campylobacter jejuni* pre-treated with noradrenaline has been shown to be more invasive in chickens.

Chronic stress leads to the spread of *C. jejuni* outside the caecum, as shown that Campylobacter does lead to diarrhoea in commercial birds. The rise of both endemic APEC and Campylobacter in broilers may be explained by the characteristic of *E. coli* to cause inflammation of the gut mucosa, making it easier for Campylobacter to invade other organs and muscle.

In conclusion, Professor Humphrey called for further investigations into improving production systems in order to reduce stress on the birds and so keep Campylobacter out. Bird type – slow versus fast growing – and environmental conditions may play a role too, he said, adding that Denmark appears to have gained better control of Campylobacter in broiler flocks through effective biosecurity.

4

Poultry Production and Health

Minnesota tops the nation in tmurkey production. The poultry industry brings in more than $800 million in income for producers, processors and other related industries. Our educational efforts translate into better bottom lines. We help the poultry industry improve management techniques, develop new products and find new markets.

This program provides information on: poultry production and processing; poultry diseases; nutrition and food safety; and poultry nutrition. We serve a broad audience including the commercial poultry industry, niche producers, consumers and youth.

Nutritional and Feeding Strategies to Minimise Nutrient Losses in Livestock Manure

The livestock industry has undergone substantial changes in the past few decades. The poultry, swine, and dairy industries in particular have become increasingly concentrated. This has resulted in fewer, but larger, livestock operations throughout the country. As a result, there has been a growing concern about odor emissions from livestock operations and potential decreased water quality caused by nutrient runoff from livestock manure. One of the key components to lowering air and water pollution from manure is the animal diet.

All animals, whether ruminant (cattle and sheep) or non-ruminant (pigs and chickens), have five basic nutritional needs. Animals need energy which is provided by fats and carbohydrates in the diet, protein, water, vitamins, and minerals. Even in the best management situations, animals are not able to utilise 100% of the nutrients that are consumed. Some nutrients will be excreted in the manure. Nutrients excreted in manure come from four primary sources: (1) feed wastage, (2) excess nutrients provided in the diet, (3) undigested nutrients in the diet, and (4) biological losses from dead cells in the body. From a manure management perspective two nutrients are of particular interest: *nitrogen* from the breakdown of protein, and the mineral *phosphorus*.

Nitrogen and Phosphorus

Nitrogen is essential in livestock diets. Proteins are made up of building blocks called amino acids, which in turn, contain the element nitrogen. Therefore, nitrogen is necessary to build protein. However, nitrogen excretion from animals is an environmental concern because it can volatilise into the air as ammonia causing offensive odours from livestock operations. Ammonia present in lakes and streams can be toxic to aquatic life. Nitrogen can also leach into ground water and cause pollution. Fifty percent of the total U.S. and 95% of rural residents obtain their drinking water from ground water sources.

Phosphorus is also necessary in animal diets. It is a key component of bone and teeth and is part of high energy compounds. Phosphorus is also an important part of cell membranes, RNA and DNA molecules. Phosphorus from animal manure binds to soil particles and can be carried by erosion to surface water such as lakes and rivers. Once in the surface water, phosphorus can cause excessive growth of algae. The algae use oxygen needed by fish, resulting in fish kills.

The average concentration of nitrogen and phosphorus in livestock and poultry manure. These are only averages. The actual concentration of nitrogen and phosphorus in the manure of a given livestock operation will vary with management, diet, manure storage facility, and age of animal housed in the facility.

Reduce Feed Waste

What can producers do to reduce the nitrogen and phosphorus that is excreted in livestock manure? One method of reducing nutrient losses in the manure is to *reduce feed wastage*. Feed wastage is any feed that is provided to, but not consumed by the animal. It can be feed that is left over in the trough or bunk or feed that is slopped over the edge of the feeder and ends up on the ground or in the pit. Van Heugten and Van Kempen (1999) found 2-12% feed wastage in the U.S. swine industry. They estimated that if just 5% of feed is wasted, a net income loss of $1.77/pig depending on the market can result and an additional 327 grams of nitrogen and 82 grams of phosphorus is excreted per pig. Producers can reduce feed wastage in their livestock operation by regularly adjusting feeders, using proper feeder design for age and type of animal, and regular maintenance of the feeders or bunks. The use of total mixed rations (TMR) for cattle and pelleted feed for swine and poultry can also reduce feed wastage.

Increase Accuracy

A second method of reducing nutrient losses in manure is to *more accurately meet the nutrient requirements of the animal.* Animals are only able to utilise a certain amount of any given nutrient for maintenance and productive purposes. Nutrients supplied above that level are excreted by the animal.

Therefore, while it is important that all animals are supplied sufficient amounts of each needed nutrient, it is also critical that producers do not grossly over-supply nitrogen and phosphorus in the diet. Male and female animals do not have the same nutrient requirements. Nutrient requirements also change as the animal progresses from one phase of production to another. For example, the nutrient needs of starter calves are not the same as the nutrient needs of feeder cattle. Therefore, animals should be grouped by gender and phase of production and fed diets specific to their nutrient needs. Feeding animals in groups according to their gender is known as split-sex feeding. Feeding according to phase of production is known as phase-feeding.

In the past, many producers added nutrients at levels above the recommended level necessary for maintenance and production as a "safety margin." Spears (1996) found that the typical diet for North Carolina finishers contained 0.40 to 0.70% total dietary phosphorus. The recommended phosphorus level for finisher pigs was only 0.40%. On average, producers were supplying their pigs with 1.55 times the phosphorus necessary for growth and maintenance.

A similar study conducted with Wisconsin dairy cattle revealed that producers were providing 0.45 to 0.50% total phosphorus in the diet when the recommended level of total phosphorus for a dairy cow was only 0.38%. Excess phosphorus provided to livestock can not be utilised by the animals and is excreted in the manure. Simply lowering the dietary phosphorus level 0.1% will decrease phosphorus excretion about 8.3% in pigs (Kornegay and Verstegen, 2001). Due to research studies such as these, many producers now lower safety margins to reduce nutrient excretion and save money.

Another way to more accurately meet the nutrient requirements of the animal is to use synthetic amino acids in poultry and swine diets. Synthetic amino acids are manufactured sources of highly concentrated specific amino acids. Traditionally, high protein ingredients such as soybean meal or meat and bone meal were used to supply the animal with necessary protein and amino acids. But these feed ingredients provide a wide variety of amino acids, not always in the specific quantities needed by the animal. When synthetic amino acids are supplemented into the diet, less high protein feed ingredients are needed to meet to the amino acid requirement of the animal, and the amino acids are added at levels more closely reflecting the nutrient needs of the animal. This reduces the total protein content of the diet and decreases nitrogen excretion in manure. For every 1% reduction in crude protein achieved by using synthetic amino acids, total nitrogen losses are reduced by 8% (Kerr and Easter, 1995). Reducing nitrogen excretion also decreases ammonia production and may lower odor emissions from the livestock facility.

Improve Digestibility

A third method of reducing nutrient losses in manure is to *improve the digestibility of the diet.* When a highly digestible diet is fed, a higher percentage of the nutrients are used by the animal for maintenance and growth and fewer nutrients are excreted in manure. Some feed ingredients are more digestible than others. For example only 14% of the phosphorus in corn is actually in a form that is available to a pig, while 90% of the phosphorus in corn distiller's dried grains with solubles is available to a pig. By selecting feed ingredients that are highly digestible, producers can reduce the nutrients excreted in manure. However, producers need to balance digestibility of the diet with economics when deciding which feed ingredients best fit into their operation.

An increasingly common way to improve phosphorus digestibility in swine and poultry diets is to add a commercially available enzyme called phytase to the diet. Most phosphorus in cereal grains fed to livestock is bound to a compound called phytate. As a result, the phosphorus is not available to the animal for growth and maintenance purposes and is excreted in the manure. Phytase is an enzyme that breaks the phytate compound and increases phosphorus availability. As a result, more of the phosphorus from cereal grains is available to the animal and less is excreted in the manure. Phosphorus excretion can be reduced 33% when phytase is added to a low phosphorus diet

(Kornegay and Verstegen, 2001). It is not necessary to add phytase to cattle diets because the microbes in their stomach naturally produce phytase. There is a variety of corn called low-phytate corn that has the same effect as adding phytase to swine and poultry diets. This corn variety has the same total phosphorus as conventional varieties, but low-phytate corn naturally has less total phosphorus bound to the phytate compound. Consequently, more phosphorus is available to the pig or chicken. Currently, it is more economical for producers to feed conventional corn with supplemental commercial phytase than to use low-phytate corn in the diet. This may change with continued improvement in plant genetics.

The implementation of nutrition and feed strategies discussed above will significantly reduce nutrient excretion, but they are not the total solution to the amount of manure produced by livestock, since 100% utilisation of a diet is not possible. As we look ahead to the future, there may be other dietary strategies that become available to assist with lowering nitrogen and phosphorus excretion in livestock manure. However, economics will always be a driving factor behind producer acceptance of these new technologies since the producer must maintain a profit to stay in business. Therefore, continued research is needed is this area to ensure that producers have economical technologies available to help lower nutrient excretion from their livestock facility.

The Effects of Plant Extracts on Bacteria by Molecular Profiling

NEXT Enhance® 150 is a standardised combination of thymol and cinnamaldehyde, an active ingredient of cinnamon extract. This combination acts to optimise gut flora equilibrium in the broiler and breeder chicken, leading to better feed efficiency and growth for the animal. Continuing to understand the mechanisms behind the consistent benefits in animal trials, Carotenoid Technologies SA (CaroTech®) has worked exclusively with the Zootechnical Department at the University of Ljubljana in Slovenia., one of the few sites in Europe which has the expertise in the state-of-the-art methodology of PCR-DGGE fingerprinting.

Previous work by the two groups has elucidated the effects of the combination of thymol and carvacrol found in Origanum spp. have been shown by molecular profiling to increase Lactobacillus populations and reduce E. coli. New scientific results have been published in Folia Microbiologica 53 (3), 204-208 (2008). Entitled *"Effect of Sodium*

Monensin and Cinnamaldehyde on the Growth and Phenotypic Characteristics of Prevotella bryantii and Prevotella ruminicola" (D. Ferme, M. Malnersic, L. Lipoglavsek, C. Kamel, G. Avgustin).

Current studies are underway with a European partner to determine the effects on Clostridium spp. and other pathogens. These novel findings not only pave the way for a better understanding of the active ingredients, but can be used as supportive data in the product registration as a zootechnical additive under European Directive 1831/2003.

Nutrition Issues Challenge Broiler Performance

Feed quality has become a formidable obstacle to achieving broiler performance goals, according to poultry nutritionists and consultants. "Our greatest challenge is ingredient quality," says Dr. Tom Frost, director of nutrition and research, Wayne Farms, Oakwood, Georgia. "We continue to receive increased pressure to use more and more byproduct ingredients, yet the quality of those ingredients continues to go down." Poultry meal is a prime example. With more of it being diverted to the pet food industry, the meal that's left is far from the quality it was before, and the same holds true for meat- and bone-meal blends and fat. "The rendering situation — it's a challenge," he says.

Another example is distillers' dried grains (DDG). It hit the industry as a prime ingredient, but the ingredients and fat content need scrutiny. "This, too, needs to be stabilised," adds Frost, who shared his views with other poultry nutritionists who came together for a roundtable discussion, "Feeding Broilers for Optimum Health, Performance and Profitability." The event was sponsored by Intervet/ Schering-Plough Animal Health.

DDG Market 'Immature'

Bill Johnson, senior nutritionist with Tyson Foods, Magee, Mississippi, says that one disadvantage to the new DDG market is that it's very immature and doesn't yet have well-established trading rules. Much DDG is traded through brokers and resellers, not by direct sale from the producers. Strong agreement came from Dr. Paul Twining, independent poultry consultant, Princess Anne, Maryland. A primary obstacle is determining the nutrient value for ingredients coming into the feed mill, and as for DDG, "you've got to have information going clear back to the production plant. There can be a tremendous difference between DDG from plant A versus plant B.

"Personally, I won't buy DDG on the open market, through a broker. I want to know exactly the plant of origin. It's the same with bakery meal," he says. Dr. Shivaram Rao, director of nutrition and feed mills for Foster Farms, Livingston, California, thinks it takes a lot of homework when alternative ingredients such as DDG are brought in. "We've got to make sure that we know the consistency and digestibility of the product so that we are not altering the amino acid balance or the calorie-to-protein ratio in the finished feed," he says. Current lab techniques are an added frustration, Rao adds, because results come back late. By the time moisture or protein changes are detected, the birds have already eaten that feed, he says.

Dr. Robert Teeter, professor of animal science, Oklahoma State University, Stillwater, says that the calorie and protein package — as well as coccidiosis control — are "extremely important" to broiler performance, but so are other aspects of the managerial environment that have a calorific cost. Lighting programs today, for instance, are saving energy on the order of about 100 kcals per kilogram. Pellet quality is another managerial factor that will increase the effective caloric value of rations.

Additional Complications

Several other factors complicate the job of the poultry nutritionist, roundtable participants said. One is extreme variability in corn quality, depending on where it was grown. "We've seen considerable differences in mycotoxin levels in the corn," Twining says. "And there may be some fairly extreme differences in the caloric content of corn that's not showing up in our standard approximate analysis."

Weather as well as housing conditions can pose additional problems that hinder broiler performance goals. "We have to make sure that the growers are on our program," he adds Dr. Mike Blair, nutritionist, Pilgrim's Pride, Woodstock, Georgia, pointed to feed mills as a limiting factor in poultry nutrition.

Bin space, especially with older mills, is a challenge and another is getting people to pay attention to how they're feeding the birds. Good communication with live service techs is imperative, he says. Looming restrictions on in-feed antibiotics is another significant challenge facing poultry nutritionists.

Although the antibiotic-free market isn't here now, Rao says it's well on the way. "In a year or two, it's going to come," he predicts. "I'm concerned that we be prepared for that — that we be a source of antibiotic-free poultry." Rao adds: "I'm kind of excited about vaccines as an alternative to some in-feed antibiotics. They're something we need to keep improving on and maybe they will give us some answers."

Lessons from Antibiotic-free Flocks

The participants acknowledged there have already been lessons learned from antibiotic-free birds that they've applied to conventional flocks. According to Twining, one lesson is that better husbandry results in fewer problems. "We've seen lower bird density in the antibiotic-free programs, and we're using some fairly high levels of probiotics in place of an antibiotic," he says. "All of these things are expensive in a large-scale production system, so it comes back to a speciality-type program where we have to have a premium to make it economical."

At Foster Farms, some management strategies from its antibiotic-free/organic program have been applied to the conventional flocks. "Mainly it's the fat quality," Rao insists. "In our antibiotic-free/organic program, we use high-quality soybean oil and the performance is better than in our conventional program. So now we have our own fat-blending facility. We have dramatically improved our fat quality."

Foster Farms has also learned that it's important to provide more space. Based on results from the antibiotic-free/organic program, "we were able to convince management to give us more space for the conventional birds," Rao adds. Occasionally, additional lines not belonging to the pattern for the zeolite appear in a pattern along with the X-ray lines characteristic of that zeolite. This is an indication that one or more additional crystalline materials are mixed with the zeolite in the sample being tested. Small changes in line positions may also occur under these conditions. Such changes in no way hinder the identification of the X-ray patterns as belonging to the zeolite.

The particular X-ray technique and/or apparatus employed, the humidity, the temperature, the orientation of the powder crystals and other variables, all of which are well known and understood to those skilled in the art of X-ray crystallography or diffraction can cause some variations in the intensities and positions of the lines. These changes, even in those few instances where they become large, pose no problem to the skilled X-ray crystallographer in establishing identities. Thus, the X-ray data given herein to identify the lattice for a zeolite, are not to exclude those materials which, due to some variable mentioned or otherwise known to those skilled in the art, fail to show all of the lines, or show a few extra ones that are permissible in the cubic system of that zeolite, or show a slight shift in position of the lines, so as to give a slightly larger or smaller lattice parameter. A simpler test described in "American Mineralogist," Vol. 28, page 545, 1943, permits a quick check of the silicon to aluminium ratio of the zeolite. According to the description of the test, zeolite minerals with a three-dimensional network that contains aluminium and silicon atoms in an atomic ratio of Al/Si=2/3=0.67, or greater, produce a gel when treated with hydrochloric acid. Zeolites having smaller aluminium to silicon ratios disintegrate in the presence of hydrochloric acid and precipitate silica. These tests were developed with natural zeolites and may vary slightly when applied to synthetic types.

U.S. Pat. No. 2,882,243 describes a process for making zeolite A comprising preparing a sodium-aluminium-silicate water mixture having an SiO.sub.2 :Al.sub.2 O.sub.3 mole ratio of from 0.5:1 to 1.5:1, and Na.sub.2 O/SiO.sub.2 mole ratio of from 0.8:1 to 3:1, and an H.sub.2 O/Na.sub.2 O mole ratio of from 35:1 to 200:1, maintaining the mixture at a temperature of from 20.degree. C. to 175.degree. C. until zeolite A is formed, and separating the zeolite A from the mother liquor. As noted above, this invention involves, inter alia, a method for reducing degradation of internal egg quality. In one of its forms, this method involves a process for obtaining an egg crop having a reduced rate of internal egg quality degradation which comprises (i) preparing a poultry diet which contains from about 0.25 to about 3.0 weight percent zeolite A, (ii) feeding such diet to laying poultry, and (iii) recovering an egg crop therefrom in which the rate in decrease in Haugh unit value of the eggs is reduced as compared to the rate of such decrease under the same conditions with the same diet absent the zeolite A.

In another of its forms, the present invention relates to a method of improving the egg shell characteristics, e.g. egg shell strength of eggs from laying poultry. A convenient means of measuring egg shell strength is by measuring the specific gravity of the egg.

The present invention is in the general field of poultry farming and relates particularly to the feeding of laying fowl or layers. In a particular aspect, it relates to a method of feeding female birds with a feed formulation which enhances, for example, interior egg quality, as well as egg production, and/or egg shell quality. This invention also relates to the eggs with improved egg shells obtained from birds fed a diet of this invention. Poultry diets of this invention are typically diets which include a zeolite such as zeolite A in an amount such that degradation of the interior quality of eggs is retarded. This invention also relates to improved poultry farming. Thus, it relates to methods which provide improved egg production. These methods comprise obtaining an increase in yield of marketable eggs from laying hens. The increase is provided by feeding hens a diet of this invention.

Description of Related Art

The literature associated with poultry nutrition, and the effect of both nutrition and egg processing on egg quality is so vast as to defy brief summation. A few of the developments in this field are highlighted below. In previously issued patents, we disclosed that zeolite A enhances egg shell quality, U.S. Pat. No. 4,556,564; improves feed utilisation efficiency, U.S. Pat. No. 4,610,882; and improves liveability, U.S. Pat. No. 4,610,883. The disclosures in those patents are incorporated by reference herein as if fully set forth. Other related art is mentioned in the patents.

An article by C. Y. Chung et al from Nongsa Sihom Young Pogo 1978, 20 (Livestock) pp. 77-83 discusses the effects of cation exchange capacity and particle size of zeolites on the growth, feed efficiency and feed materials utilisability of broilers or broiling size chickens. Supplementing the feed of the broilers with naturally occurring zeolites, such as clinoptilolite, some increase in body weight gain was determined. Chung et al also reported that earlier results at the Livestock Experiment Station (1974, 1975, 1976—Suweon, Korea) showed that no significant difference was observed when 1.5, 3, and 4.5 percent zeolite was added to chicken layer diets.

U.S. Pat. No. 3,836,676 issued to Chukei Komakine in 1974 discloses the use of zeolites and ferrous sulfate crystals in an odourless chicken feed comprising such crystals and chicken droppings. The results were said to be no less than those in the case where chickens were raised with ordinary feed. Experiments have been in progress in Japan since 1965 on the use of natural zeolite minerals as dietary supplements for poultry, swine and cattle. Significant increases in body weight per unit of feed consumed and in the general health of the animals was reported (Minato, Hideo, Koatsugasu 5:536, 1968). Reductions in malodor were also noted.

Using clinoptilolite and mordenite from northern Japan, Onagi, T. (Rept. Yamagata Stock Raising Inst. 7, 1966) found that Leghorn chickens required less food and water and gained as much weight in a two-week trial as birds receiving a control diet. No adverse effects on health or mortality were noted. The foregoing Japanese experiments were reported by F. A. Mumpton and P. H. Fishman in the Journal of Animal Science, Vol. 45, No. 5 (1977) pp. 1188-1203. Canadian Pat. No. 939,186 issued to White et al in 1974 discloses the use of zeolites having exchangeable cations as a feed component in the feeding of urea or biuret non-protein (NPR) compounds to ruminants, such as cattle, sheep and goats. Natural and synthetic as well as crystalline and non-crystalline zeolites are disclosed. Zeolites tested included natural zeolites, chabazite and clinoptilolite and synthetic zeolites X, Y, F, J, M, Z, and A. Zeolite F was by far the most outstanding and zeolite A was substantially ineffective.

In a study at the University of Georgia, both broilers and layers were fed small amounts (about 2%) of clinoptilolite, a naturally occurring zeolite from Tilden, Tex. The egg shells from the hens receiving zeolite were slightly more flexible as measured by deformation, slightly less strong as measured by Instron breaking strength, and had a slightly lower specific gravity. The differences in egg shell quality were very small. This type of zeolite was ineffective in producing a stronger egg shell. An article written by Larry Vest and John Shutze entitled "The Influence of Feeding Zeolites to Poultry Under Field Conditions" summarising the studies was presented at Zeo-Agriculture '82. A study by H. S. Nakone of feeding White Leghorn layers clinoptilolite, reported in 1981 Poultry Science 60:944-949, disclosed no significant differences in egg shell strength between hens receiving the zeolite and hens not receiving the zeolite.

Background of the Invention

As is well known, the interior quality of eggs is diminished with time. The loss in quality attributes of the albumen and yoke is a function of temperature and movement of carbon dioxide through the shell. Low temperatures decrease the rate of loss in Haugh units—a standard measure of interior egg quality—and thus it is recommended that eggs be stored at temperatures close to the freezing point, a procedure which, as a practical matter, is not always feasible. To reduce rate of carbon dioxide (and moisture) loss various shell treatments have been utilised, such as spraying oil on the eggs.

A way has been discovered for reducing, if not eliminating, the need for refrigeration and spray oiling on eggs as a means of preserving interior egg quality. To ameliorate the problem of interior egg degradation with time, the following procedures have been recommended heretofore:

1. Gather eggs three to four times per day.
2. Clean the eggs promptly after gathering and cool for 12-24 hours at 13.degree. C. or preferably 10.degree. C. before packing in cases or cartons.
3. Keep the eggs at between 60 to 85%, preferably 70 to 80%, relative humidity.
4. Resort to careful handling.
5. Use proper packing using precooled containers only.
6. Resort to frequent marketing of not less than twice a week.
7. Use speedy, refrigerated transportation and make frequent deliveries to sales outlets, preferably at least five times per week.
8. Use adequately refrigerated holding spaces at sales outlets.
9. Keep the eggs in home refrigerators at 7.degree. C. to 13.degree. C., and preferably use all of the eggs within one week.

Summary of the Invention

The demand for poultry eggs, especially chicken eggs expanded considerably over the last decade. The poultry industry has grown from a home industry to a large scale manufacturing industry in which tens of thousands of eggs are produced daily at single farms or egg laying installations. Some eggs are produced for eating and some eggs are produced for hatching. One problem with such large scale egg producing is premature reduction in interior egg quality as a function of time.

That is, unless eggs are handled and/or treated in accordance with the above 9-point program, their quality as regards internal qualities of the albumen and yolk may deteriorate faster than would be desired. Moreover, a way of simplifying egg production, distribution, marketing, etc. with concomittant reduction in refrigeration requirements would be a welcome contribution to the art. Another problem associated with large scale egg production is breakage. Even a slight crack in an egg makes it unsuitable for hatching and most other marketing purposes. It is estimated that some six percent of all eggs produced are lost for marketing because of cracking or breakage. Shell strength is very important to inhibit breakage. The stronger the egg shell, the less likely the egg will be cracked or broken. Machinery and techniques necessary for carefully handling the eggs to avoid breakage are expensive and time consuming.

Another substantial loss of egg production, estimated to be about a seven percent loss, is the production of shell-less eggs. Any reduction in shell-less eggs can be an important factor in large scale egg production.

We have discovered that internal egg quality can be maintained at a higher level by feeding laying poultry a diet containing zeolite A. In other words, this invention provides, in one of its embodiments, a process for reducing the rate of internal egg quality degradation which comprises feeding a laying poultry hen a diet which contains zeolite A such that the rate in decrease in Haugh unit value of the eggs from said hen is reduced as compared to the rate of such decrease under the same conditions with the same diet absent the zeolite A. In this process, the amount of zeolite A is usually within the range of about 0.25 to about 3.5 weight percent, preferably between about 0.75 to about 1.5 weight percent.

As stated above, our previously issued U.S. Pat. No. 4,556,564 discloses that zeolite A in poultry feed improves egg shell quality.

In Nutrient Requirements of Poultry, Eighth Revised Edition 1984, National Academy Press, Washington, D.C. (1984) it is state, that leghorn-type chickens, both layers and breeders, have a dietary requirement of 0.15 weight % chlorine per day. Assuming an average daily intake per hen of 110 g of feed per day, this amounts to 165 mg of chlorine, per hen per day. This requirement is usually furnished as chloride ion, typically in salt, i.e. NaCl.

In the practice of this invention, the chloride level of the poultry diet may be reduced in order to improve egg yolk production and/or egg shell quality.

At the same time, the utilisation of such a diet achieves improved retention of internal egg quality as manifested by egg crops in which the rate in decrease in Haugh unit value of the eggs is reduced as compared to the rate of such decrease under the same conditions with the same diet absent the zeolite A.

Accordingly, we have discovered that egg production, egg quality, notably internal egg quality, and/or egg shell quality of poultry eggs is enhanced if, besides adding a zeolite (notably zeolite A) to the diet, the chloride level in poultry diets is decreased below the previously recommended requirement. The improvement can result in the production of more collectable eggs, which with broiler breeder hens means more eggs that are settable in incubators. The enhancement of results provided by addition of zeolite A is unexpected. Furthermore, reduction of chloride levels below the previously required amount is unexpected.

Description of Preferred Embodiments

This invention comprises a method for improving the internal quality of eggs in terms of preservation of Haugh units. This method involves, in another embodiment, a process for obtaining an egg crop having a reduced rate of internal egg quality degradation which comprises (i) feeding laying poultry a diet which contains from about 0.25 to about 3.0 weight percent zeolite A, and (ii) recovering an egg crop therefrom in which the rate in decrease in Haugh unit value of the eggs is reduced as compared to the rate of such decrease under the same conditions with the same diet absent the zeolite A. In addition, this invention comprises a method for improving eggs, egg production and/or egg shell quality.

This method comprises feeding birds a diet which (a) contains reduced amounts of chlorine, e.g. less chloride than what was previously recognised as required, and which (b) also contains a zeolite such as zeolite A. Thus, in an important aspect, this invention in one of its forms can be considered a method of enhancing the quality of eggs, both in interior quality and in egg shell quality by use of a zeolite. In still another preferred embodiment, this invention provides an egg produced by laying poultry fed a diet which contains zeolite A, the egg being characterised by exhibiting a rate of decrease in Haugh unit value that is lower than the rate of such decrease under the same conditions with the same diet absent the zeolite A. Preferably, these eggs are produced by a chicken hen, although the invention is deemed applicable to comparable improvements in duck eggs, quail eggs, turkey eggs, and the like.

Zeolites are crystalline, hydrated aluminosilicates of alkali and alkaline earth cations, having infinite, three-dimensional structures.

Zeolites consist basically of a three-dimensional framework of SiO.sub.4 and AlO.sub.4 tetrahedra. The tetrahedra are cross-linked by the sharing of oxygen atoms so that the ratio of oxygen atoms to the total of the aluminium and silicon atoms is equal to two or O/(Al+Si)=2. The electrovalence of each tetrahedra containing aluminium is balanced by the inclusion in the crystal of a cation, for example, a sodium ion. This balance may be expressed by the formula Al/Na=1. The spaces between the tetrahedra are occupied by water molecules prior to dehydration. Zeolite A may be distinguished from other zeolites and silicates on the basis of their composition and X-ray powder diffraction patterns and certain physical characteristics. The X-ray patterns for these zeolites are described below. The composition and density are among the characteristics which have been found to be important in identifying these zeolites. The basic formula for all crystalline sodium zeolites may be represented as follows:

Na.sub.2 O.Al.sub.2 O.sub.3.xSiO.sub.2.yH.sub.2 O

In general, a particular crystalline zeolite will have values for "x" and "y" that fall in a definite range. The value "x" for a particular zeolite will vary somewhat since the aluminium atoms and the silicon atoms occupy essentially equivalent positions in the lattice. Minor variations in the relative number of these atoms do not significantly alter the crystal structure or physical properties of the zeolite. For zeolite A, the "x" value normally falls within the range 1.85.+-.0.5.

The value for "y" is not necessarily an invariant for all samples of zeolites. This is true because various exchangeable ions are of different size, and, since there is no major change in the crystal lattice dimensions upon ion exchange, the space available in the pores of the zeolite to accomodation. This is simply done by immersing the egg in solutions of salt water of varying strengths. It is well known in the art that specific gravity correlates with egg shell strength. As specific gravity of the egg is raised, the strength of the egg shell is increased.

This latter method of this invention can also improve egg shell thickness, and/or decrease the number of eggs produced without shells. The improvements of this invention are obtained by adding a zeolite to the diet of laying poultry. The zeolite can be added to regular or standard poultry feed or administered directly to the birds by some other means.

The percent of zeolite (such as zeolite A) in the diet is typically from about 0.25 to about 3 or 4 weight percent, as discussed below. For this invention, chloride intake of the birds is reduced, even below the levels previously recognised as required. For leghorn-type poultry the chlorine level in the diets used in this invention are below 0.15 weight percent; more preferably, between about 0.04 and 0.10 weight percent. The benefits of this invention can be obtained with ETHACAL.RTM. feed component, a commercially available form of sodium zeolite A. As indicated above, Zeolite A is added to such feed formulation in small amounts by weight percent of up to about four weight percent. Greater amounts may be used, but may deprive the poultry of the desired amount of nutrients. Greater amounts are also likely to be cost ineffective. A preferred amount of zeolite A is from about one-half to about two percent by weight of the total feed formulation. A most preferred amount of zeolite A is from about 0.75 to about 1.50 weight percent of the total feed formulation. For this invention the term "poultry" includes all domestic fowl, namely chickens, turkeys, ducks, geese, and the Corn is the principal diet for most laying poultry. A feed formulation comprising by weight percent the following is desirable

Summary of Results

Examples 1 and 2 demonstrate quite clearly that use of zeolite A in poultry feed improves the resistance of the eggs to internal quality degradation over time. Besides enhancing egg quality during storage and transportation, these results are of importance in that the longer hatching eggs are stored prior to incubation, the lower will be the percentage of hatch.

Thus, the invention provides eggs with superior characteristics both as regards "freshness" for human consumption, and enhanced hatchability. Indeed, some published studies indicate that chicks hatched from eggs stored for over 8-10 days perform more poorly when fed to market weights. Thus, improvements in these characteristics should likewise be achieved by utilisation of this invention.

In Example 4 hens fed the zeolite A, low chloride diet laid eggs with the highest specific gravity, shell thickness and percent shell, and they produced the lowest percentage of shell-less or membrane eggs, Hens fed the zeolite A diets containing high chloride, equivalent to the chloride level of the control diet (Diet 1), also laid eggs with the same improvements, but of lesser magnitude than the low chloride fed hens.

These improvements result in the production of more collectable eggs, which with broiler breeder hens means more eggs that are settable in incubators. That inclusion of zeolite A in the diets have no adverse effect on overall egg production or feed consumption. That inclusion of zeolite in the diet results in increased weight gain suggesting that improved feed utilisation is occurring.

The above results suggest the use of zeolite A levels in the diet of from about 0.25 to about 3.0 weight percent, or up to about 4.0 weight percent. The above results also suggest that for chickens, both layers and broiler breeders, the chlorine content (% chloride) in the diet can be less than about 0.15 weight percent. Preferably the chloride content is from about 0.04 to about 0.10 weight percent.

The results set forth in the Examples above demonstrate the advantages of this invention in improving interior egg quality, egg shell quality, shell thickness, percent shell, or in reducing the number of shell-less or membrane eggs. Any such benefit other than improved interior egg quality, for the purposes of this invention is referred to as an improved egg shell characteristic.

The above results demonstrate an improved process for enhancing either or both of (i) interior egg quality and (ii) an egg shell characteristic which comprises regularly feeding to laying poultry a diet which comprises a zeolite such as zeolite A. To achieve the benefits of (ii) the diet also contains a reduced chlorine content. As shown, the diets with reduced chlorine content can contain an amount of chlorine (as chloride) which is below that which was previously recognised as required. Such diets are referred to herein as chloride deficient diets. These diets can comprise the chloride content supplied by the corn and soybean meal or similar portions of the diet without any added chloride source (such as salt, i.e. NaCl), or with an amount of salt or other chloride below that which is commonly added to the diet, as illustrated above.

Generally speaking, one uses enough zeolite such as zeolite A to provide the benefit or benefits conferred by such additive. Preferred ranges are set forth above. Besides processes comprising feeding birds, this invention also provides chloride deficient diets or feed formulations which have zeolite and chloride contents as discussed above. Preferred aspects of this invention comprise use of zeolite A diets (including chloride deficient diets of zeolite A) for poultry, although diets of this invention can be used in other aspects of animal husbandry such as in the care and feeding of endangered, wild species. Of the poultry species, chickens, including layers and broiler breeders are highly preferred. Use of the diets of this invention provides improved methods of poultry farming which can result in more eggs that are usable, and which have desirable characteristics.

Having described this invention and its benefits in detail above, it will be apparent that those skilled in the art can make modifications and changes of the invention as above-described without departing from the spirit and scope of the claims which follow.

Contributions from/by Livestock-keepers

Subsidies can become a mechanism for securing farmers' involvement in trials and treatments that they do not really consider to be worthwhile: for this reason they should be avoided or minimised. On the other hand, there are potential problems in avoiding them altogether from the outset. First, this may make it difficult for resource-poor livestock-keepers to participate in trials, particularly where the treatment has to be purchased. Second, many rural people have a dependency mentality, having become accustomed to receiving government handouts, and hence are reluctant to pay for things themselves when they are working with development agencies. This is particularly the case in India.

Unless the livestock-keepers have a strong sense of ownership of the trial, and believe that it is very important and has a reasonable chance of success and profitability, they are unlikely to pay for the whole treatment themselves, particularly if it is one that they are not very familiar with. This is illustrated by the experience of a de-worming trial in Kenya, The treatments used in the trials conducted by the BAIF/NRI project were subsidised to varying degrees.

The basis for this was that the technologies were new to the goat-keepers, and that they were therefore taking a risk (financial and potentially to the health of their goats) in applying them. Urea Molasses Granules was the newest of all the treatments, so a 100% grant was given for this.

In this kind of situation the researchers should develop a clear understanding with the livestock-keepers that the size of the subsidy will be reduced, year by year, as the goat-keepers become familiar with the technologies and see the benefits they have on their animals (assuming, that is, that they are effective!).

Incentives for Participants in Control Groups

For most kinds of trials it is desirable to have a group of animals that do not receive any treatment, so that comparisons can be made between the two groups to see what difference the treatment made. The non-treatment group is known as the 'control' group.

If the owners of the control group animals also own the treatment group animals, motivation is not likely to be a problem. However, if they are different to those of the treatment group motivating the control group owners to participate actively in the control group can be a challenge. They may be reluctant to participate, because they do not see themselves or their animals gaining anything from their involvement in the trial.

If this situation arises, it should be pointed out to them that if the technology on trial proves to be effective they too will benefit after the trial has been completed. If that is not enough of an incentive, however, they can be offered some kind of material incentive. For example, in the BAIF/NRI project control group goat-keepers participating in a trial focusing on milk yield were given metal drink containers; and in other trials a breeding buck was made available to both the treatment and control groups.

Apparently, ICRAF do not carry out on-farm feeding trials comparing treatments and a control, only different treatments, partly because of the issues raised (Morton, 2001). We would argue strongly in favour of having a control group, and finding ways of motivating them, such as those just mentioned. They are just as important to a trial as the treatment group members, so researchers should put just as much effort into developing their sense of ownership and involve them fully in group meetings, etc.. If there is a need to hire the services of a local monitor for monitoring the trial, preference can be given to someone from the control group.

Human Resources

Researchers' attitudes are very important. They should have respect for the views and knowledge of the people with whom they intend to work. They should be prepared, for example, to traverse difficult terrain and climates, leave home early and return late, spend time in the project area, and hold interviews at times that are convenient for local people.

This is particularly important when working with women. FPR/PTD requires staff skills that are often scarce. Many researchers in national agricultural research organisations may lack experience and aptitude in working with farmers in a participatory way. While many NGO staff may have little, if any, experience The precise nature of the disciplines will depend to some extent on the constraints and opportunities that livestock-keepers identify as being most important. For example, if heavy worm burdens appears to be a serious problem, a parasitologist may be required to identify which helminths are involved and to advise on an effective treatment. Alternatively, if the focus were on improved forage production, then a forage agronomist and livestock nutritionist might be needed. However, the basic minimum should be one livestock scientist and one social scientist or agricultural economist.

Ideally, at least one member of the research team should have had some previous experience of participatory research. In addition, it is generally desirable, and sometimes essential, to have at least one woman in the team, to facilitate interaction with female livestock-keepers. There is often a need for training of research staff in participatory research methods. To some extent it may be possible to provide 'on-the-job' training, but if funds permit short (2-day to 2-week) formal courses would be ideal..)

Financial Resources

The financial resources required for collaborative research are often underestimated. Monitoring and analysis may be more time-consuming than they are in more conventional research modes. With on-farm experimentation there is generally more variability in the experimental data than is the case with the contract mode, and consequently data may need to be obtained from a larger number of fields, farmers, etc., if the effect of the technology being tested is to be detected. It may also be necessary to carry out more detailed monitoring of farmers' management activities to explain some of the variability in results between herds. It will also take time, and hence money, to build-up an effective working relationship.

Research projects normally have a fixed duration, which is usually no more than 3 years. Research scientists need to consider whether FPR can deliver/meet objectives in the time earmarked for the project, while bearing in mind the need to (a) allow for delays and complications, and (b) develop a rapport and partnership with participants. For example, if they have not worked closely with local communities already a long lead-in time may be required to develop the necessary rapport to work effectively with them.

Any type of livestock research, whether participatory or not, should be based on a sound understanding of what livestock-keepers see as their priority needs - their main constraints or opportunities. In the case of PTD this is an essential foundation for the whole process, since people are unlikely to invest time and effort on research that they consider to be unimportant. Obtaining an accurate understanding of needs and priorities can be difficult and time-consuming, and may require at least two phases of discussions with farmers. Directly asking people their most pressing problems may merely generate well-known 'shopping lists'. Problems are likely to be described in terms of a lack of an input (which the farmer hopes the project will provide): for example, where there is a high mortality rate, livestock-keepers may characterise the problem as a "lack of veterinary medicines". It is important to identify the underlying cause of the problem, rather than just the symptoms.

Getting farmers to rank problems or priorities, starting with the most serious or important, provides more information than simply making a list; it also reduces the risk of researchers distorting farmers' views to fit in with their own personal interests or priorities. A companion guide to this one, entitled Participatory Situation Analysis with Livestock-Keepers: A Guide (Conroy, 2001), provides valuable guidance on conducting needs assessments with livestock-keepers, including the use of problem-trees to identify underlying causes. Researchers should seek to increase their understanding of what is required and possible as the research progresses.

Ideally, the researchers assist livestock-keepers in tackling their most pressing production problems. However, not all of these problems can be easily solved by improved technologies (although there may still be scope for influencing policies or institutions) and their research institute may not have the expertise (or resources and willingness) to address certain problems. Nevertheless, so long as the issue is a reasonably high priority for the local people, and researchers are also convinced, there is likely to be potential for fruitful collaboration.

The broader the scope of the project, and the greater the variability of the systems and situations of the target group(s), the more time and effort will be required for the needs assessment exercise.

Scheduling

The timetable for the research needs careful consideration. There are often conflicting demands regarding the pace at which research progresses. The short duration of many research projects (often no more than 3 years) may encourage researchers to establish on-farm trials as quickly as possible, so that they can gather data for a minimum number of seasons or years.

On the other hand, it is desirable to avoid rushing because this can undermine the participatory approach as farmers/users are not given an equal say in the design of the research, and they may not develop a sense of ownership. It may be desirable to focus on interacting with fewer farmers and villages during the first season or year so that staff become familiar with a participatory approach. This is a time when effective relationships are developed with participants, research opportunities are identified, and the technologies to address them are agreed. It is important that the research team members are not over-ambitious and do not spread themselves too thinly. A narrowly focused approach will be preferable where:

- resources are limited;
- staff are relatively new to PTD; or
- a lot of time needs to be invested in building up rapport with the communities.

Where the approved duration of the research is longer (say up to 5 years), it may be best to have an initial phase of observing farmers' own informal trials or current practices, in relation to selected issues. This helps the researchers understand problems and potential solutions from the farmers' viewpoint, gain deeper insight into the differing problems of individual farmers or sub-groups, and prepares them for participatory research if they are new to it.

Observations of this kind can be conducted concurrently with exploratory trials, in which the researcher contributes ideas.

In many cases a preliminary needs assessment (taking a few weeks or months) is followed quickly by exploratory experiments. Through monitoring these experiments and discussion at the end of the season, the process of assessing, and reassessing, needs continues.

This works if researchers allow themselves enough time for quality interaction with farmers, carry out genuinely exploratory experiments, maintain an open mind on the problems, and do not insist on repeating the early experiments over several seasons in order to obtain 'conclusive' data.

Identifying Where to Work and with Whom

Selecting Farming Systems and Zones

In determining the areas and production systems in which the participatory research will be conducted, both biophysical and socio-economic factors are relevant. Secondary data sources should be consulted, and full use made of whatever information is already available so that duplication is avoided. In many countries maps already exist that identify the various agro-ecological zones: information on the spatial distribution of different livestock species and ethnic groups can be obtained from census data. Some of the information required may, however, have to be collected by the project through short overview surveys, to enable characterisation of farming systems in a way that is most relevant to the project.

Initial characterisation can be modified as further information is collected during the course of a project, and if time is pressing target groups can be developed iteratively, during needs assessment, monitoring and on-farm experimentation. Pastoralist is defined here as a household in which at least two-thirds of its income comes from sale of livestock and livestock products or livestock-related activities. Thus, their livestock production systems are commercial by definition. Semi-commercial means that the sale of animals to meet contingencies (i.e. use as liquid assets) and the sale of animals as a profit-making enterprise are both important.

If a project is oriented towards a particular commodity (e.g. buffalo), it may adopt a different approach to targeting from one which is oriented to a particular area or category of farmers. It may, for example, be appropriate to choose an area that is known to be important for the livestock species concerned. In the case of the BAIF/NRI project, the researchers were aware that the main purpose and benefits of goat-keeping vary, and production systems vary accordingly from one area and group to another. Although some of the problems encountered are broadly similar, the most appropriate technologies for addressing them may also vary.

Thus, we chose to work in districts that between them represented a range of different production systems. Initially, the project was only working in Rajasthan (with small and marginal farmers) and Gujarat (with landless or near landless pastoralists). We also wanted to work with landless labour households, in which one or more members were involved in agricultural wage labour. This type of goat-keeper was not very common where we were working initially, so we expanded the project coverage into Maharashtra and Karnataka, where they constituted a larger proportion of the rural population. We also wanted to work more with women, and we knew where there were landless women goat-keepers in Maharashtra.

Multi-locational or multi-production system trials have a further benefit. Scientists gain a clearer picture of production variability in the on-farm trials (from location to location or production system to production system), so they are in a better position to judge the situations and locations (recommendation domains) in which the new technology could be successfully applied (Waters-Bayer, 1989).

Selection of Research Locations

Practical considerations will inevitably limit the choice of specific research locations – 'villages', 'communities', or perhaps a network of local specialists.

The further away these locations are from the researchers' base(s), and the greater the distance between the participating farmers, the greater the costs in time and fuel, and the less contact there is likely to be between participants and researchers.

Trade-offs may be necessary between the extent to which the locations included are representative, and the resource costs involved. A guiding principle is that well-informed choices are always preferable to the selection, by default, of non-representative situations (e.g. adjacent to research stations, major roads, previous projects, a researcher's home village, etc.).

Where researchers or collaborating organisations have already been working with certain villages for some time, and have developed a good rapport with community members, this may be a strong reason for selecting such villages in preference to others, provided they are reasonably representative of villages in the area concerned.

This can save time and resources in that a good rapport with participants already exists, and the project may easily access valuable secondary data about livelihood systems, social and economic composition and problems and priorities.

Selection of type of Livestock

In cases where a project is working in a particular geographical area, and is not tied to any particular livestock species, the researchers will need to decide which livestock species they are going to prioritise. Selection criteria that should be considered are: the potential for research to address livestock-keepers' (not researchers') priority problems associated with the most important type(s) of livestock; where research is likely to produce the greatest benefit, taking account of the seriousness of the problem (e.g. size of mortality rate) and the number of animals experiencing it; and what kind of research is likely to provide the greatest benefit to the poorest groups.

Identifying Participant Livestock-keepers

Options for engaging participants include: (a) volunteering (as individuals or community representatives); (b) delegation of selection to the community; (c) probability sampling (for a discussion of conventional approaches to sampling in agricultural projects); (d) guided purposive selection. Researchers have tended to take a somewhat ad hoc approach, and/or to favour options (a) or (b), on the basis that they are more participatory than (c) and (d).

Approaches (a) and (b) tend to bias the selection, skewing participation away from the poorest, for two reasons. First, within communities power is distributed unevenly and often volunteer or community-nominated participants are male and resource richer. Second, for many of the poorest a prolonged involvement in research activities is not attractive, as they are preoccupied with more pressing livelihood issues. FPR projects need to engage in more systematic selection strategies if they want participants to be generally representative or from particular socio-economic groups. The selection procedure needs to be discussed with the collaborators, and agreement reached on criteria and objectives.

Purposive rather than completely random selection is likely to be the most feasible approach. Purposive selection requires a prior understanding of the socio-economic composition of the village or community and inter-household relations so that farmers' views and reactions can be seen and understood in context; the project should seek to improve its understanding of the local social structure as it progresses.

Techniques such as wealth-ranking and social mapping, if used with skill and sensitivity, can provide the kind of information that is needed initially; and secondary data should be utilised when available. If the relevant information is not available when the initial participant selection process takes place, and volunteer or delegated sampling is used, the project should subsequently check the characteristics of the participants against those of the community as a whole.

Additional participants can then be selected if necessary, to make the sample more representative of the target group. Research objectives are also likely to influence the type of collaborator required. If the research is focused on a particular type of livestock, or a particular kind of livestock production system, then the participants will have to be people who meet these criteria. For example, in a scavenging poultry project working in Tamil Nadu's Namakkal District, three categories of production system were identified, and each category is likely to be represented roughly equally when trials begin.

If the research is testing a new technology the project staff may decide that it is necessary to select willing risk-bearing participants with more resources (e.g. land, labour, equipment) and/or previous positive experience in technology innovation. For programmes that run for a long time, there is a question of whether to continue collaborating with the same small group of livestock keepers, or to change every so often. Generally, this issue has to be looked at in relation to research objectives, and to the importance of maintaining rapport and relations with the community. In practice, it is likely to be expedient to maintain contact with some of the more interested farmers over a period of years, and also allow space for new farmers to join in as others decide to drop out or as new opportunities arise as the experimental programme expands.

Selection of Experimental Animals

The type of animal selected will be determined to a large extent by the nature of the trial. For example, if the experiment is focusing on weight gain in young animals the animals selected will have to be from the relevant age group; whereas, if it were focusing on conception rates they would have to be mature females. Within each of these categories, however, there is scope for varying degrees of precision. This relates to the problem of inter-animal variability. In order to minimise variability the animals should be as similar as possible.

In one of the first trials conducted by the BAIF/NRI project, which focused on young goats, the age spread of the young goats was quite large, creating unnecessary variability and making the use of a standard treatment for all of them questionable. In a similar trial the following year the age of the goats selected was more homogeneous. To minimise inter-animal variability in trials that are focusing on milk production, it may be necessary to select animals that are all of the same lactation, and perhaps ones that are all at a similar point in the lactation cycle.

Another consideration is that where animals in the treatment group belong to different people from those in the control group, the animals in each group should belong to many different owners. Otherwise, the practices of someone owning a large number of animals could become confounded with the comparison between treatment and control groups. For example, in the BAIF/NRI project's first Bhilwara trial, 13 of the 25 goats in the treatment group were owned by one person. Thus, although the treatment group does produced more kids than those in the control group, the difference could have been due to this one goat-keeper having superior goats or feeding practices, rather than to the treatment itself.

Identifying Interventions

In the early stages, the aim of the discussions with farmers is to reach agreement on the research agenda. Deciding on the research agenda is a process which should be based on an adequate understanding of the local farming system, including interactions between various components and enterprises in the system, and who is involved in, decides on and benefits from the various activities.

This understanding will help to reduce a long list of possible experiments, to one or two which are most useful and likely to bear fruit. While dialogue between researchers and farmers in this process is essential, dialogue with other knowledgeable researchers may also be vital, in order to avoid duplication and unproductive experiment-ation.

If a consultative mode is used within a national agricultural research system (NARS) setting, further discussions and consultation with other specialists on the extent of the problem and what can be done about it may be required after the needs assessment. Within a community-oriented collaborative mode, this is a joint process between the local people and the researchers.

Whatever the mode, to sustain a credible partnership with farmers and other stakeholders, the probable relevance of possible interventions needs to be gauged through careful study and widespread consultation. If researchers are convinced, but the collaborating farmers are reluctant, it may be worth organising a farmer tour to visit an area where this technology is being practised, or to a research station, before trying to introduce it in an on-farm experiment. The researcher should try and avoid the temptation to tell the local people what to do early on in the discussions. Ideas for interventions may come from any of three general sources:

- members of the local communities
- other livestock-keepers or NR users in the region
- researchers (and extensionists), based on their own organisation's work or the general body of scientific knowledge.

The local people should be encouraged to develop their own ideas initially. It may be useful to discuss ways in which group members have already tried to tackle the problem previously identified, and what effect this had. Discussions should also screen indigenous technical knowledge and previous experimentation by villagers. Often there are recognised specialists within or near a community, and it may be worth identifying these and inviting them to join in discussions, or making visits to them later for more in-depth discussions.

For example, in Rajasthan there are local specialists in animal health care in many villages, called Gunis. In recent years there have been numerous initiatives in different parts of India (by NGOs, KVKs etc.) to document ethnoveterinary knowledge and practices, so it would be wise to check whether there are any relevant studies available in your area or nearby.

Identifying Interventions – Some Examples

Members of the local communities In a feed supplementation trial in Dharwad, Karnataka, the idea for the treatment came from one of the goat-keepers. The treatment was a mixture of sorghum and horsegram. Other livestock-keepers in the region In a de-worming trial in the same district the idea for the treatment came from the practice of another ethnic group from a nearby area, who keep buffaloes. The treatment was the trichomes (hairs) from the pods of a leguminous creeper, mixed with jaggery, a lukewarm sugary solution. Researchers or extensionists In a supplementation trial in Bhavnagar, Gujarat, the researchers suggested the use of Urea Molasses Granules, which they had recently tested in a pilot project elsewhere in Gujarat.

An example from the DAREP project, Kenya Mange had been identified by goat-keepers as a major problem. Farmers considered commercial products for treating it to be too expensive, and had started looking for locally available alternatives. The project staff sought to identify alternatives by holding a number of group discussions and by visiting local herbalists, and came up with a list of eight local concoctions already being tried by farmers. The list was further screened through discussion with farmers, and a trial was designed comparing three of the local treatments with two of the recommended commercial ones and a herbal treatment of Neem solution.

Understanding the Nature of the Problem

This issue is discussed in the companion volume to this one, where tools for determining priority needs are described (Conroy,

2001). These include: (a) participatory problem tree analysis, which livestock-keepers can use to separate the core problem from its causes and effects; and (b) the participatory herd history method, which is a quick method for generating reasonably reliable baseline data on kidding rates, mortality rates etc. Understanding the nature of a problem correctly requires researchers to know what the objectives of the livestock-keepers are.

Sometimes the livestock-keepers (or the researchers) may not know, or be sure of, the nature of the problem; but they may have an informed hunch as to the cause. In these situations it is essential for both parties to discuss and explain their ideas and hypotheses before selecting treatments. One common problem that is sometimes poorly understood by goat-keepers is that of high worm burdens, which can result in high mortality rates in kids due to a combination of factors. A high worm burden in the mother may lead to a dramatic reduction in milk production, thus weakening the kid; and if the kid itself then becomes infected it will have a poor chance of survival (Peacock, 1996). However, the goat-keepers may not realise that helminths are causing the death of their kids, nor understand the life cycle of helminths and their effect on the goat's physiology. If so, the researchers will need to educate the livestock-keepers on "the underlying principles of the technology effect" (Mason et al., 1999).

This is similar to the situation with integrated pest management of crops, where 'farmer field schools' have been developed as an educational tool (Ooi, 1998). The high importance livestock-keepers attach to visible problems may also mean that they prefer to take a curative approach (i.e. respond to the problem when it appears) rather than a preventative one. This was found to be the case in one de-worming experiment on cattle and sheep (Mulira et al, 1999).

5

Breeder Nutrition and Chick Quality

The developing embryo and the hatched chick are completely dependent for their growth and development on nutrients deposited in the egg. Consequently the physiological status of the chick at hatching is greatly influenced by the nutrition of the breeder hen which will influence chick size, vigour and the immune status of the chick.

Table: The necessary change in hatchery or broiler performance to equalise profitability when breeder feed cost is changed by 1% per tonne (for example from £UK 140.00/tonne to £UK141.40/tonne or £UK138.60/tonne).

Hatch of total eggs (%)	0.24
42 day liveweight (g)	7.4
42 day FCR	0.0015
42 day mortality (%)	0.07-0.45*

**depending on age of mortality. Calculated using input-output values for UK industry 2003 (Kemp and Kenny 2003).*

The Financial Effects

Nutritional decisions for breeders need to take account of the overall economics of the whole production cycle. Table shows the changes in hatchery and broiler performance that are required to equalise the effect of a 1% increase in breeder feed cost on the profitability of the whole production cycle. Only one of these changes is required to have the necessary economic effect; in practice all are likely to move positively making the measurements of any one change difficult. The calculations are done under typical UK 2003 conditions and they show quite clearly that small improvements in bird performance are required to 'pay' for more expensive breeder feed. Conversely, apparent savings in breeder feed cost can readily lead to an overall loss if small changes in broiler performances are ignored.

Similar economic analyses have been conducted by Mississippi State University which, based on US integration 2002 costs, demonstrates that a measurable improvement in progeny liveability as a result of hen diet change can be profitable. The key point is that trying to cut the cost of a breeder feed may easily reduce the profitability of the overall enterprise.

Influence of Feed Allocation

Underfeeding the hen can have an impact on chick quality and this is particularly noticeable in the early production period. Modern hybrid parent flocks commence production at a faster rate than in the past and consequently egg output increases over a shorter time span during the early laying period.

Feed allocations during this period have not necessarily increased in line with this egg production trend. Low feed allocation intake by young commercial breeder flocks has been shown to compromise nutrient transfer to the egg, resulting in increased late embryonic death, poorer chick viability and uniformity.

In a recent study by Leeson (2004) broiler breeders were fed different levels of feed through peak production varying from 140 to 175 grams. Although the increased feed allocation increased bodyweight there was no influence on egg size, however chick weight was influenced by feed allocation. Of equal importance is the effect of overfeeding on ovarian development.

In experimental studies ad libitum feeding has been the most widely used model for overfeeding which can result in excessive follicular development or Erratic Oviposition and Defective Egg Syndrome (EODES). Flocks with EODES generally have poor shell quality, a reduced duration of fertility and poor hatchability. It is also known that fewer sperm will survive but it is not clear how the surviving sperm are affected and if they generate a weaker embryo. The same authors also warn that the effect of aggressive feeding two to four weeks after photostimulation reduces productive performance throughout the life of the flock.

In this period the bird switches from primarily growth to a reproductive state. The young birds' reproductive hormone system is not mature enough to deal with high nutrient intakes; nutrients are instead metabolised to egg yolk lipid which contributes to excess follicle development.

Research shows that nutrient supply to the broiler breeder is of consequence to chick quality and production performance.

This places greater emphasis on the nutritionist providing the correct nutrient density diet and the flock manager to provide appropriate feed intake to the bird coming into lay.

Table: *The effects of breeder feed levels on chick weight.*

Peak breeder feed (g/b/d)	***30 week breeder chick weight (g)***
140	40.3
147	40.0
155	41.5
162	41.7
169	41.8
175	42.0

Diluted Breeder Diets

The use of diluted breeder diets is receiving a lot of attention in Western Europe on the basis of improvements in bird welfare.

Experimental work feeding low energy density diets to young parent stock gave a delayed onset of oviduct development, increased early egg size, faster development of the embryo and a higher live weight of day old chicks.

When broiler mortality was above average, low density broiler breeder feeds gave a significant reduction in mortality of offspring. Other experimental work showed improvements in breeder productive performance when diluted diets were fed in the rearing period.

Vitamins

Vitamins are involved in most metabolic processes and are an integral part of foetal development, therefore the consequence of suboptimal levels of these nutrients in commercial diets are known to result in negative responses to both parent and offspring performance. Vitamins account for about 4% of the cost of a breeder feed, so economising on vitamin inclusion rates is rarely an option

Table: *Some practical recommendations for vitamin supplementation of breeder feeds*

Vitamin	*Leeson & Summers (1997)*	*DSM*	*Ross*
A (iu/g)	7	10-14	13
D3 (iu/g)	3	2.5-3.0	3
E (mg/kg)	25	40-80	100
K (mg/kg)	3	2-4	5
Thiamine (mg/kg)	2.2	2-3	3
Riboflavin (mg/kg)	10	8-12	12
Pyridoxine (mg/kg)	2.5	4-6	6
B12 (mg/kg)	0.013	0.02-0.04	0.03
Nicotinic acid (mg/kg)	40	30-60	50
D-pantothenic acid (mg/kg)	14	12-15	12
Biotin (mg/kg)	0.2	0.2-0.4	0.3
Folic acid (mg/kg)	1	1.5-2.5	2

Generally there is a shortage of information on vitamin requirements of broiler breeders especially when related to offspring performance. Most of the breeder work is quite dated and since that time breeder performance has changed. It would be impossible to review all the literature in this article, however a review of work on fat soluble vitamins, biotin and pantothenic acid have shown that vitamin E has the largest impact on progeny.

Table: *Impact of dietary breeder vitamin status on bodyweight, enzyme activities, tissue characteristics and immunity of progeny.*

Vitamin	*Progeny response*
Vitamin A	Increased liver vitamin A in embryonic and chick liver but decreased vitamin E, carotenoids and ascorbic acid. Surai et al. (1998).
Carotenoids	No positive impact on chick growth, organ development or humoral immunity in chicks five weeks post-hatching. Haq et al. (1995).
Carotenoids	Transferred from the hen to the yolk but not absorbed well by the embryo and subsequent chick. Haq and Bailey (1996).
Carotenoids and Vitamin E	Carotene, vitamin E, and their combination improved and vitamin E lymphocyte proliferation, but only vitamin E improved humoral immunity. Haq et al. (1996).

Contd...

Vitamin	***Progeny response***
Vitamin E	Vitamin E levels of 150 and 450mg/kg increased passively transferred antibody levels in chicks to Brucella abortus up to seven days of age. Jackson et al. (1978).
Vitamin E	Increased vitamin E in chicks' yolk sac membrane, liver, brain and lung all of which had reduced susceptibility to peroxidation. Surai et al. (1999).
Vitamin E	Increased progeny antibody titers to sheep red blood cells at hatch. Boa-Amponsem et al. (2001).
Vitamin E and Selenium	Increased liver glutathione activity in chicks. Increasing selenium increased selenium dependant glutathione peroxidase in chick liver. Surai (2000).
Vitamin D	Tibial calcium was increased at two weeks post-hatching and tibial ash increased at four weeks of age by increased vitamin D3. Ameenudin et al. (1986).
Vitamin K	Chicks from hens fed vitamin K deficient diet had reduced tibial glutamic acid levels at day one and 28 post-hatching but tibial glutamic acid was restored by supplementing the chick diet with vitamin K. Lavelle et al. (1994).
Biotin	Foot pad dermatitis and incidence of breast blisters were decreased in some trials in chicks from hens fed biotin fortified diet. Harms et al. (1976).
Biotin	As biotin increased in the hens' diet, yolk and chick plasma also increased. Biotin concentration in chick plasma was poorest from young hens. Whitehead (1984).
Pantothenic acid	Liveability of chicks was best when hens were fed 20mg/kg diet of pantothenic acid. Utno and Klieste (1971).

Adapted from M. Kidd 2002

The production and economic effects of vitamin E supplementation are best shown by Hossain et al (1998) where a basal corn soya feed was supplemented with 25, 50, 75 and 100mg/kg vitamin E. The effects on hatchability were not significant; however the best hatchability was obtained at 50mg/kg at 52 weeks. Offspring immune response continued to increase up to 100mg/kg. In the same studies higher final bodyweights at 42 days, improved FCR and reduced mortality were observed in chicks from eggs which had been injected with vitamin E in ovo.

Haq et al., (1996) working with very high levels of vitamin E (134mg/kg versus 412mg/kg) found no growth response to 21 days and an improvement in FCR for the offspring of hens receiving the supplemental feed. In other studies the combination of selenium and vitamin E to broiler breeders has been shown to increase liver glutathione activity of progeny. In general it seems to be justified to supplement practical breeder feeds with 100mg/kg vitamin E. There appear to be mixed reports on the efficacy of vitamin C; some experiments suggest a positive response, but a more recent study failed to detect any benefit on any production parameter.

This lengthy study used corn soya diets supplemented with 75mg/kg stabilised vitamin C which when analysed recovered 49mg/kg which might explain the variability of response.

The influence of increased vitamin levels fed to young parent stock on progeny performance is an area which has received significant commercial interest. Work conducted at Aviagen Ltd has shown chicks derived from 31 week old parent stock fed elevated levels of vitamins showed improved growth to 11 days and reduced mortality compared to chicks derived from 42 and 45 week old parents.

Similar responses have been found in the field where chicks derived from young parents fed increased levels of vitamins have benefited in terms of viability and liveability.

Perhaps this supports the need for further work exploring the vitamin requirements of the breeder in the early production period.

Table: *Blood cell count of the broiler derived from parents fed high or low vitamin and mineral levels*

	Breeder low vitamins/minerals	***Breeder high vitamins/minerals***
Heterophil	5.3	3.8
Lymphocyte	4.6[a]	21.4[b]
Monocyte	1.1	5.3
Basophil	0.0[a]	5.4[b]

Whitehead (1991) proposes that a basis for making recommendations is to feed vitamin levels that maximise the resulting level in the egg. For vitamins with active transport mechanisms (thiamine, riboflavin, biotin, cobalamin, retinol and cholecalciferol) these levels reflect the saturation of binding proteins.

Levels derived in this way include 10mg/kg for riboflavin and 250 microgram/ kg for biotin.

Whitehead (1991) contrasts this level of riboflavin with the conventional requirement (4mg/kg in this case) but the higher figure – the upper limit to nutritionally useful range – may be a better guide to good commercial practice.

Table: *Summary of minerals fed to breeders shown to have an effect on progeny performance.*

	Growth	*Liveability*	*Immune function*	*Skeletal*
Fluoride				X
Phosphorus				X
Selenium		X		
Selenomethionine	X		X	
Zinc	X		X	X
Zinc and methionine		X		

Adapted from M. Kidd 2002

Vitamins and Chick Immunity

Reference has already been made to the effect of vitamin E on chick health and immune function, while other vitamins have been researched none show the same degree of effect as vitamin E. Recent work by Rebel et al (2004) investigated the effects of several elevated levels of vitamins and trace elements fed to breeders and broilers on the immune system of birds infected with malabsorption syndrome.

Broilers derived from breeders fed elevated vitamins and mineral levels had increased numbers of leukocytes at day old which indicated stimulation of the immune system.

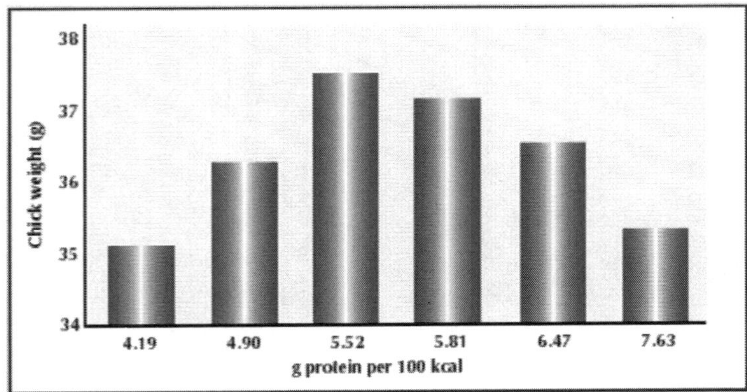

Figure: *The effect of protein-to-energy ratio in the breeder feed on chick weight at hatch*

Major Minerals

Calcium, phosphorus, sodium, potassium, magnesium and chloride are involved in shell formation hence general improvements in shell quality lead to better egg and chick quality. Variations in maternal phosphorus supply have been shown to influence bone ash of young but not older progeny. Broiler performance was not affected by these treatments so the practical significance of this work is not clear but the use of relatively low phosphorus levels in breeder diets, while benefiting egg shell quality, may not lead to the best possible bone integrity in the early stages of growth.

Trace Minerals

Most interest in this field has centred on the use of chelated minerals which have been shown to increase deposition in the egg and transfer to the tissues of the hen and the embryo. Most recent work has focused on the antioxidant status of breeders, embryos, offspring and the role of selenium. Surai (2000) has shown the role of Selenomethionine on both vitamin E and glutathione peroxidase levels in eggs, embryos and chicks up to 10 days of age.

The economic benefits of using Selenomethionine compared with sodium selenite have been examined in a number of unpublished field trials in the UK. Hatchability improvements ranged between 0.5-2.0 chicks per 100 eggs and in another trial 0.3-0.7 chicks per 100 fertile eggs. Few of these tests involve a proper assessment of subsequent broiler performance although comments about chick quality are generally positive.

In one of the commercial trials mentioned an improvement of 0.5% in mortality and cull rate at 10 days was observed when organic selenium replaced sodium selenite. Research has indicated that the improvements in chick immunity as a result of mineral fortification of hen diets may result in improved liveability. Flinchum et al. (1989) demonstrated that leghorn breeders fed supplemental zinc methionine to a zinc adequate diet had progeny with improved survival to an E. coli challenge. Similar improvements to progeny liveability were seen with breeders fed supplemental zinc and manganese amino acid complexes.

Nutrient Levels in the Breeder Diet

There is clear evidence that a high protein to energy ratio depresses hatchability, and probably chick performance. The experiment by

Whitehead et al. (1985) shows the effect of excess protein where the higher protein level reduced reproductive performance, producing 3.1 fewer chicks per 100 fertile eggs.

Chick quality was also reduced so that the difference in saleable chicks was 4.0 per 100 fertile eggs. The effect of energy protein ratio in the breeder feed is shown in figure.

This emphasises both the effects of excess and inadequate protein, and also indicates that the optimum level is quite steeply defined. According to this trial the optimum protein level is at 5.52g protein per 100kcal which converts to an optimum of 15.18% protein for a diet containing 2,750cal/kg of feed. The protein level of the diet and its ratio to energy is important not only for parent performance but also for chick quality.

Table: *Commercial comparison of breeder feeds based on wheat or maize (400g/kg)*

Advantage of maize over	*wheat based feed*
Mortality during lay (%)	-1.7
Total eggs (per hen housed)	+3.8
Hatching eggs (per hen housed)	+4.8
Hatching/total eggs (%)	+0.9
Hatch of set eggs (%)	+0.6
Hatch of fertile eggs (%)	+1.1
Second quality chicks	-0.1

Based on a comparison of two commercial houses each containing 6500 female grandparent breeders. Data to 58 weeks (Ross Breeders, unpublished data, 1998).

The Effect of Feed Ingredients

There is evidence of improved breeder performance when maize is compared to wheat as the main cereal in breeder feeds. From a survey of many depleted commercial flocks overall hatch of fertile eggs in the UK based on wheat diets and Brazil based on maize diets is 83.3 and 86.2 per 100 eggs respectively.

Other management factors may contribute to this difference in hatchability other than cereal source; male management is very good in Brazil and the resulting high fertility may also contribute something to this difference. Unpublished commercial development trials from the Netherlands and Aviagen Ltd grandparent flocks support this observation.

The most likely benefit of maize is probably in shell quality and thickness. From the same data average poorer shells with specific gravity of <1.08 accounted for 26.1% of eggs from wheat fed hens and 17.1% from maize fed. Studies of hatching losses showed less late dead embryos (>18 days) and less bacterial contamination.

These two responses are expected with eggs of better shell quality. Evidence about fat levels and sources is conflicting but there is no question that this is an important consideration.

Added fat levels should be kept low in breeder feed (1-3%) and preference given to unsaturated vegetable oils rather than saturated animal fats.

Work from Mississippi State University compared maize oil and poultry fat and generally supported the use of more unsaturated fat. Maize oil increased 21 day bodyweight over that of poultry fat and improved slaughter weight of broilers in comparison to equal levels of poultry fat and lard.

Table: *Experiments comparing fat sources and/or levels for broiler breeders.*

Reference	***Fats compared***
Brake (1990)	PF
Brake et al. (1989)	PF
Denbow & Hulet (1995)	SBO, PF, FO
Peebles et al. (1999a, b)	CO, PF, LA
Peebles et al. (2000a)	PF, CO, LA
Peebles et al. (2000b)	PF, CO, LA

Fats: PF – poultry fat; SBO – soybean oil; FO – fish oil; CO – corn oil; LA – lard

Over and undersupply of nutrients into and through lay can have a very significant impact on breeder production and quality of progeny. This places greater emphasis on the nutritionist providing the correct nutrient density diet and the flock manager to provide appropriate feed allocation in lay.

Addition of micronutrients to the breeder has been shown to be beneficial to progeny quality especially in the early production period. Use of specific dietary ingredients such as maize can affect breeder performance and progeny quality. Both on economic grounds and on biological grounds, high quality nutrition of breeders is well justified.

Broiler Breeders - Feeding Breeders to Optimise Chick Quality

Integration concentrates the majority of their energy and technical resources on broiler production. Breeders generally do not receive the attention they deserve. Integrators are looking to maximise egg and chick production at the lowest cost with insufficient attention being given to the quality of the chick. There is an abundance of evidence that confirms that breeder nutrition directly impacts the progeny. However, these impacts are generally difficult to measure under commercial conditions and even under controlled trial conditions they may be obscured. The obsession with lowering feed cost has, I believe, led us to a point where some nutrients, like vitamins and trace minerals are marginal in modern breeders. While, others like crude protein are generally fed in considerable excess and this profoundly affects breeder performance and health.

Due to fierce competition and the subsequent low profit margins, vitamin producers are reluctant to invest resources in research, so little breeder vitamin and trace mineral research has been conducted over the past 15 to 20 years. In addition, the current benchmarking practices have pressured the industry in lowering cost wherever possible. Fifteen years ago, breeders were generally peak fed 165 to 170 grams per bird per day with peak egg production at 65 to 70 per cent. Today, the majority of breeders receive 145 to 150 grams of feed at peak with peak production at 80 to 90 per cent – this on reduced vitamin and trace mineral levels.

Measuring breeder nutrition impact on progeny is difficult due to a number of factors. Firstly, the assumption is that all birds receive the allocated feed amount; this certainly is not the case with 30- to 40-gram differences between birds being common. The birds that will be nutrient-limiting are those that are the most severely restricted. This may only be five per cent of the flock, so when assessing the commercial performance between two nutritional regimens it is difficult to have a large enough sample to be able to measure the impact. Secondly, some of the effects may only be apparent under stress conditions. Small or no differences may be seen under trial or ideal commercial conditions, but under a stress (disease, low broiler nutrient levels, chilling), the responses will be more apparent.

The Nutritionist's Role in Breeder Nutrition

Since breeders are restrict fed, it is the nutritionists responsibility to design a feed that meets the nutrient requirements of the bird under their particular set of conditions of; egg production, environmental temperature, etc. and then to ensure that that the highest quality ingredients are used in these feeds and that the feeds remain constant over time. Since the birds are restrict fed, the breeder manager needs to 'learn' how to feed the birds. With constantly changing feeds this makes every flock a new experiment, which makes it difficult to manage body weights and uniformity. Birds do not understand percentages.

Producers focus on the nutrient composition of the feed. The most common question I hear is, "What crude protein and energy level should I be feeding?" We need to understand the daily nutrient needs of the bird for a particular environment, given amount of activity and egg production. We then need to know what feeding system is being used and how well feed is distributed to the birds. Once we have this information we know daily need for: kcals of ME, mg of amino acid, grams of calcium etc that the bird requires and how many grams we need to feed it. Then we can formulate the feed to meet these requirements.

Energy and Fat

Energy is a fuel for maintaining basic metabolism, mobility and egg production. Energy is not a nutrient as such but is the result of metabolic oxidation of energy yielding nutrients – carbohydrates, fats and proteins. Having reviewed a large number of breeder formulations, energy appears to be the fist limiting 'nutrient'. This is true under both hot and temperate environments. Under hot conditions fat is helpful as a source of readily available energy, to help with the increased energy demand for increased rate of respiration – panting. Under more temperate conditions, carbohydrates are the preferred source of energy as they are not as readily metabolised and do generate additional heat during metabolism (heat increment). Fat composition and quality is important for the breeder and progeny. Essential free fatty acids are required for cell membrane integrity, immune competence, fertility and embryonic development.

It is well documented that maternal feed fat composition directly influences the fat composition of the egg. This includes fatty acid profile and fat soluble vitamins. Wang et al. (2002) were able to show that maternal dietary polyunsaturated fatty acids (PUFAs) affected the PUFA levels in the spleen of progeny. They also demonstrated feeding Omega 3 fatty acids to the breeder resulted in higher Omega

3 fatty acid levels and immune responses in progeny. The immune response was measured by reduced wing swelling in birds injected with bovine serum albumin. Some other effects of vitamin supplementation will be addressed under the vitamin section.

Crude Protein and Amino Acids

Crude protein is a measurement of the nitrogen content of a feed or an ingredient, assuming that amino acids are the source of measured nitrogen. Crude protein does not give us any information about the amino acid composition and or availability. Modern nutritionists should only formulate feeds on digestible amino acid levels, using crude protein solely as means of easily analysing blended feeds to ensure correct blending.

Commercial feeds are generally still formulated to minimum crude protein levels, which results in feeds that, with the exception of TSAA (total sulphur amino acids), have amino acids significantly higher than required. The critical amino acids in breeder nutrition are methionine (TSAA), tryptophan, lysine and isoleucine. Methionine and tryptophan directly impact egg size and egg production. A corn/ soy diet has excess lysine and isoleucine. When formulating to minimum crude protein levels, the lysine levels are often up to 40 per cent above requirement. Work by Lopez and Leeson (1995) clearly illustrated the negative effect of excess crude protein on fertility and DeBeer and Coon (2006) identified lysine and isoleucine as two amino acids directly affecting the fertility of breeders.

Minerals

All nutritionists are aware of the critical importance of the macro minerals; calcium, available phosphorus, etc. on bone and egg shell integrity. However, there is very good evidence that maternal levels of trace minerals especially zinc, manganese, copper and selenium impact levels in the egg and influence progeny. This is not new research; hen feeds deficient in zinc were shown by Edwards et al, (1959) to cause slow growth of chicks; Turk et al. (1959) to cause weak chicks, poor feathering and high mortality; and Kidd et al. (1992) showed supplementing with inorganic and/ or organic Zn increased levels of Zn in the bones and increased bone weight.

In addition to Zn's role in DNA and enzymes, Zn is particularly important in the young bird in the synthesis of two key proteins: collagen and keratin (Underwood and Suttle, 2001). Keratin is a structural protein in skin and feathers. Collagen is the major structural protein of internal tissues, including cartilage and bone. The cost of ensuring a sufficient and an available Zn source in breeder and early broiler feeds is minimal considering the impact of poor skeletal development and compromised immunity on the profitability of broiler production. As with Zn, Mn, Cu and Se levels can be affected in the egg by maternal supplementation. Manganese is vital in embryonic and post-natal bone development. Cu is essential for reproduction and development. Se has a sparing effect on vitamin E as an antioxidant.

Vitamins

Although severe vitamin deficiencies will cause a wide variety of deformities and severely affect hatch, grossly underfeeding vitamins is not commonly seen in practice. It is the marginal deficiencies caused either by low supplementation, sources of questionable quality/ availability and less dominant breeders consuming less than calculated feed quantities. The progeny will not exhibit classical deficiency syndromes, but they will not perform to their potential. Aviagen conducted a study to asses the impact on progeny where Vitamins E, K and B vitamins were supplemented at 20 per cent above Aviagen breeder recommendation. Although broiler body weights were only 20 grams heavier at term, mortality of the supplemented group was 2.2 per cent lower than the control birds, with a yield advantage of 0.2 per cent at 2kg body weight.

Egg yolk vitamin E levels were measured and a 50 per cent increase in α-tocopherol seen. In addition, candling clears at 60 and 64 weeks of age were 12.2 vs. 17.3 and 17.9 vs. 26.9 per cent per cent for the higher supplementation group. Adequate vitamin and trace mineral supplementation with quality vitamins and available minerals is an inexpensive way to ensure that the young chick is prepared for optimal skeletal growth and a healthy immune system, to help deal with challenges during brooding.

Amino Acids and Enteritis in Broiler Chickens

An intact healthy gut is vital to the health and welfare of young chickens. The gut as a multi-cellular organ; has absorptive enterocytes, goblet cells for mucin production; immune cells and an intrinsic neural system. The gut is the first line of defence against, dietary toxins and enteric pathogen invasion. According to Reed (2001) more than 65 per cent of dietary threonine is used in gut function; the threonine content of mucin is excess of 30 per cent. Enteritis results in considerable loss of mucin, studies have shown that threonine will be partitioned away from protein deposition to support the gut and mucin production. To demonstrate the importance of threonine under enteric stress, a threonine dose response study was conducted to evaluate the impact of L-threonine on enteritis in broiler chickens. Birds were reared to 14 days on a coccidiostat-free, corn/soy feed with increasing levels of added L-threonine (at lysine to threonine ratios from 64 to 72). At 14 days, the birds were challenged with *Clostridium perfringens* and grown to 42 days of age. Body weights of the challenged birds at the higher threonine levels showed a positive response at 31 days and were similar at 42 days. Breast meat yield on the high treatment was similar between the challenged and control groups at 42 days.

This study supports the theory that animals will partition threonine away from muscle protein to support gut health and mucin production (Ajinomoto Heartland 2003). Breeder nutrition directly affects the health, egg production and fertility of the birds. There is considerable evidence that breeder nutrition affects the progeny. Although we seldom see the gross deformities of severe nutrient deficiencies, it would appear that current nutrient levels, particularly trace mineral and certain vitamins are marginal particularly for those less aggressive birds that are unable to consume their expected feed allotment. It is these subtle deficiencies that could benefit from proper supplementation of these nutrients.

Overfeeding of crude protein:

1. Costs money; a recent evaluation indicated around $3.00 per ton.
2. Results in additional heat increment, which contributes to the heat load under heat stress conditions. New high yield breeds are particular sensitive to heat stress and nutritionists need to help minimise this stress.
3. Negatively affects fertility
4. Excess protein is excreted, and elevates the nitrogen in the faeces. High nitrogen in the litter results in foot pad lesions and excess ammonia, which can irritate mucus membranes of the eyes a respiratory tract.

Breeder hens are producing the broiler chicks that are the basis of our business. Feeding them correctly will result in healthy, virile chicks, which will have the immune system and skeletal framework required to deal with challenges and achieve their growth potential. The increase in cost of feeding a breeder hen properly compared to the total cost of producing a mature broiler chicken is marginal. Saving cents may be costing you Rands.

Energy Costs Associated with Commercial Broiler Production

Data recently compiled at the Applied Broiler Research Unit may be of value in assessing your farms' energy demand and (based on your costs for fuel and electricity) monetary expense to meet this demand.

Housing and Management Practices

The information presented represents data from 38 consecutive flocks of straight run broiler chickens grown at the Applied Broiler Research Unit during the period October 1996 through June 2003. All flocks were grown for the same integrator under a standard broiler production contract.

The houses were all 40 × 400 ft. Two houses (1 and 3) featured conventional cross-ventilation with low-pressure foggers, while the other two houses were curtain-sided and tunnel ventilated. One tunnel ventilated house (4) had evaporative cooling pads and the other (2) had an experimental sprinkler system.

Detailed descriptions of the houses, environmental control systems, sprinkler system, and housing modifications was given by Berry et al. (1991), Xin et al. (1993) and Tabler and Berry (2001). Management practices were the same in all houses and the farm manger was the same individual throughout the study period. Half of the 38 flocks were grown for 49 days or less while the other half were grown for more than 49 days. The youngest flock was 39 days at harvest while the oldest was harvested at 57 days.

Propane Usage

The lower fuel consumption in House 3 during the winters of 1998 and 1999 was due to use of an experimental wood-burning pellet furnace. House 4 was the most challenging house to control from a management standpoint since it had more ammonia than any other house. This increased ammonia required increased ventilation to maintain the proper environment resulting in increased gas consumption during cooler periods of the year.

Although House 4 consumed the most fuel during the seven-year period, it should be noted that when the 1998 and 1999 data were ignored, tunnel ventilated houses consumed only 2% more fuel than did conventional houses. Also, if the ammonia problem in house 4 could be solved, tunnel ventilated houses would likely consume less fuel than conventional houses.

Electricity Usage

The summer of 2000 was the most costly in terms of electricity usage followed by the summer of 2001. However, unlike propane usage, each house accounted for an equal amount of electricity usage (25%) during the 7-year period. The increased electrical demand in House 3 during the winters of 1998 and 1999 are again associated with use of the wood-burning pellet furnace in that house.

Even though House 1 showed the highest kWh usage as compared to the other houses, there was less than 185 kWh difference between houses and the houses were the same when compared on a percentage-of-use basis. Electricity usage for the farm over the 7-yr period averaged 12,617 kWh per flock. Based on 5.5 flocks per year the farm would have used 69,394 kWh per year or 17,348 kWh per house per year. If electricity costs $0.06 cents per kWh, electricity costs would come to $4,164 for the farm or $1041 per house.

Total Energy Costs

However, because fuel costs are much greater than electricity, growers have a much more serious problem dealing with high fuel bills in the winter than they do the electric bill in the summer. Contract growers face numerous challenges associated with raising broilers. One significant challenge is the monetary expense related to fuel and electricity costs. Energy data from 38 consecutive flocks of straight run broilers over a seven-year period at the Applied Broiler Energy Unit indicate that approximately 25% of the gross farm income is required to pay the annual propane and electricity bills and that propane costs may be roughly four to five times the cost of electricity. While energy costs will vary somewhat from farm-to-farm, the wise use of energy should be a priority for all growers.

Nutrient Digestibility of Broiler Feeds Containing Different Levels of Variously Processed Rice Bran Stored for Different Periods

Nutrient digestibility of broiler feeds containing different levels of variously processed rice bran stored for varying periods was determined. A total of 444 Hubbard male chicks were used to conduct four trials.

Each trial was carried out on 111 chicks to determine digestibility of 36 different feeds. Chicks of 5 wk age were fed feeds containing raw, roasted, and extruded rice bran treated with antioxidant, Bianox Dry (0, 125, 250 g/ton), stored for a periods of 0, 4, 8, and 12 mo and used at levels of 0, 10, 20, and 30% in feeds. Digestibility coefficients for fat and fibre of feeds were determined. Increasing storage periods of rice bran significantly reduced the fat digestibility of feed, whereas no difference in fibre digestibility was observed. Processing of rice bran by extrusion cooking significantly increased digestibility of fat even used at higher levels in broiler feeds. Interaction of storage, processing, and levels was significant for fat digestibility. Treatments of rice bran by different levels of antioxidant had no effect on digestibility of fat and fibre when incorporated in broiler feed.

Meeting the Nutrient Requirements of Broiler Breeders

Major changes have taken place with the feeding of broiler breeders in recent years. One of the most significant of these changes was separate sex feeding. With such an approach to feeding it is now possible to more closely control the weight of the pullet coming into production and to be confident that, with challenge feeding, the additional feed is being used to maximise egg output rather than put extra weight on already overweight cockerels. By being able to precisely control cockerel weight, marked improvements in hatch of total eggs set, with reduced cockerel numbers, is being attained. With separate sex feeding, changes in ration formulation to more precisely meet nutrient requirements of males and females are now possible. Also more attention can now be paid to flock uniformity and by switching to everyday feeding, uniform pullet flocks that reach peak production in 4 to 5 weeks, and produce 180 to 190 eggs per breeder, is not uncommon. While there are many points to consider with respect to management, feeding, health care, photoperiod, etc., that can have a bearing on the economics of production, the present article will deal mainly with feeding to more precisely meet the energy and protein requirements of the modern day broiler breeder.

Body Weight and Condition

While body weight recommendations are readily available for the various commercial breeders on the market today, often there is little information given on how to keep problem flocks close to these suggested weights. A 6 week old commercial female broiler will weigh around 1800g while a weight of 600g is the target weight for a broiler breeder pullet of the same age.

Hence, it is evident that overweight, not underweight pullets are usually the problem during the growing period. Pullets that are overweight during rearing should not have their feed allotment reduced in order to bring their weight in line with suggested target weights. Rather, feed should be held at the same level until body weight comes in line with that recommended for a given age.

Then adjustments in feed allowance should be made to hold pullets to the desired weight curve. Another important point is to ensure that pullets, as they approach mature pullet weight, are carrying sufficient body flesh.

This can be observed by checking fleshing over the keel. Pullets that are lacking in flesh are immature birds and will be delayed coming into production. Increase feed, if necessary, to ensure well fleshed pullets and delay light stimulation of the flock until proper body condition has been achieved, even though body weight may be slightly in excess of the weight guide.

Nutrient Requirements

Two of the major and most costly nutrients required by the breeder are energy and protein. While protein requirements have often been the major consideration of most producers, usually it is energy, the most costly dietary nutrient, which has the most influence on performance.

Energy

Energy is often referred to as "The Fire of Life". Feed supplies energy, which in turn is required for growth, egg production, maintenance, vital life functions and activity. In developing feeding programmes for broiler breeders consideration should be given to body size, as this is a major factor, along with environmental temperature, influencing the maintenance energy requirement. Production of a product is the other major consideration. This may include a growth component but the main consideration is egg mass output.

Maintenance

The maintenance energy requirement is affected by body size, environmental temperature and level of activity. In order to calculate energy requirements a knowledge of the calorie requirement per unit body size (surface area), growth rate, and egg mass output must be known. Environmental temperature must also be taken into account as it will affect the energy needs for maintenance.

This is important for as will be discussed, the energy requirement for maintenance makes up by far the greatest percentage of the birds total energy requirement. A number of equations have been developed which give reasonable estimates of the total energy requirement of a broiler breeder. However, as with any predicted estimate, such values must be evaluated for each and every situation.

Energy Requirements for Pullets

It is important that growers recognise that the major factor influencing feed intake is the bird's need to meet it's maintenance energy requirement. Since environmental temperature is the main factor influencing maintenance energy requirement (other than body weight), pen temperature should be closely monitored and feed intake adjusted accordingly if significant temperature changes result.

While precise changes in feed allotment, to account for changes in environmental temperature, are difficult to predict accurately, a reasonable estimate would be that for every 1°C change in pen temperature through a temperature range of 15 to 30°C, the bird will alter its feed intake by 3g/d. However at high or low temperatures the change in feed intake would be significantly greater.

Energy Requirement for Broiler Breeders

This decreases noticeably at peak production and continues to peak egg mass when the hen would be partitioning a significant proportion of its' energy intake to meet its requirement for egg mass output. Again if one is to precisely meet the breeder's energy requirement, factors influencing this requirement must be considered. One is the increase in body weight and this is apparent by the steady increase in maintenance energy requirement as the production cycle progresses. However, total energy requirement peaks around peak egg mass production and then declines as the requirement for egg mass production is reduced as egg production decreases.

It is important that the producer recognise the fact that only around 20% of the energy intake of the hen is diverted into egg production and increased body weight (which would be fairly small). Since the hen will preferentially partition nutrients to meet her requirement for maintenance, if her feed allowance is not sufficient to meet her total energy requirement, egg mass output and thus egg production and/or egg size will be reduced.

Influence of Feather Condition or Energy Requirements

While pen temperature has a significant effect on maintenance energy requirement, the feather condition of the hen can significantly influence the affect of environmental temperature. A marked increase in feed intake is noted with birds as they lose feathers, which often occurs with aging as well as poorly managed flocks. Allowance must be made for feather condition when estimating the calorie needs of a flock if maximum production and egg size are to be maintained.

Reducing Feed Intake Past Peak Production

It is common practice to reduce feed allowance to a flock shortly after peak production is attained. This may be responsible for some of the dips in production noted shortly after peak is attained. Another error that a number of producers make is to reduce feed allowance in relation to the drop in egg production, (eg.) if production drops 5%, feed allowance is reduced by a similar amount.

An average egg of around 65g requires approximately 140kcal of metabolised energy for its production. If feed allowance is 160g/b/d at peak production, with a diet containing 2800 kcal/kg, this will result in a daily intake of 448 kcal ME/b/d. If 140 kcal are required for the production of one egg this means that approximately 31% of the hen's energy intake is going towards egg production at this time. Egg mass output at 85% production and with a 65g egg = 55.3 g/h/d. If production drops to 80%, and during this time egg weights increases by 1g, egg mass output would be 52.8g/h/d. Thus egg mass output would drop;

$$\frac{55.3 \div 52.8}{55.3} \times 100 = 4.5\%$$

Hence, the energy required per hen per day for egg production should drop;

$$\frac{4.5 \times 140}{100} = 6.3\text{kcal}$$

Thus for a 4.5% drop in egg mass output total dietary energy requirement has only dropped

$$\frac{6.3}{448} \times 100 = 1.4\%$$

...(not the 5% that one might estimate from the drop in egg production). It should be obvious that decreased feed intake after peak production has to be precisely calculated as the hen's requirement is not related directly to the drop in egg mass output. From the above discussion it should be apparent that there is a strong possibility that the broiler breeder is deficient in energy intake up to and perhaps beyond peak egg mass production. As already mentioned any deficiency in energy intake will readily translate into reduced egg numbers or smaller eggs or both.

Protein: Since the maintenance requirement for energy is so high, in relation to the breeders total energy requirement, energy intake is one of the main factors influencing the reproductive performance of broiler breeders. Thus changes are often made in daily feed allotment in order to try and maintain an optimum intake of energy. However, little attention is paid to the amount of protein consumed by the hen. Most breeder diets contain between 16 and 18% protein. With variable feed intakes during the laying period, daily protein intake obviously varies. Many breeders are fed up to 160g of feed per day (or more). Thus protein intake could range from 25.6 to 28.8g per day, and in many cases higher).

Protein Required for Egg Production

What is the calculated protein requirement for the broiler breeder? Using values generated for commercial layers, the protein required to produce a 65g egg (containing 7.8g of protein) should be around 7.8÷.55 (suggested efficiency of dietary protein utilisation for egg production) = 14.2g.

Protein Required for Maintenance

With an estimated total endogenous loss of nitrogen, including feather loss, to be 280 mg/kg of body weight to the 0.75 power, a 3.5kg hen would require 3.5^{75} X 280 = 717mg of nitrogen per day to meet her maintenance requirement. Converting this to protein required would give (.717 X 6.25) or 4.48g of protein. Assuming the hen is 55% efficient in converting dietary protein to body proteins the hen would require 4.48 ÷ .55 = 8.15g of protein intake per day to meet her maintenance protein requirement.

Total Daily Protein Requirement of a Broiler Breeder

Considering egg production and maintenance the protein requirement would be 14.2 + 8.15 = 22.4g/h/d. This should be sufficient for a hen to lay a 65g egg every day. While no allowance has been made for body weight gain, this should be minimal after peak production and besides much of this gain will be fat deposition and thus a minimum of body protein would be deposited. Since every bird in a flock is not laying an egg a day the protein required for egg production per day is only the average percent production times the estimated

14.2g of protein required for the 65g egg. Thus the daily protein required of a flock of broiler breeders would be significantly less than 22.4 g/b/d. While the above values are only estimates of protein requirements, they are reasonably good estimates based on values generated mainly from egg production hen values.

Partitioning the Breeders Protein Requirement

While it was shown that around 80% of the energy consumed is partitioned to meeting the hens requirement for maintenance (including weight gain and activity) it can be estimated from the values generated for protein requirements that, (8.15 ÷ 14.2 X 100) approximately 57% of the protein intake of the breeder is going to meet its' requirement for egg mass production. Thus the main factor influencing the protein requirement of the broiler breeder is egg mass output, not body maintenance as with energy.

Estimated Protein Requirements for Flock Breeders

Percent Dietary Protein to Meet Requirements: This would translate into dietary protein levels of from 7 to 11.4% to meet these estimated requirements, assuming an average feed intake of 160g/b/d. While a lot of assumptions and estimates have been made in generating the above values the low levels of dietary protein suggested are not too far removed from the estimates suggested by Bowmaker and Gous (1989), Harms and Ivey (1992) and Lopez and Leeson 1993).

As can be noted in 10% dietary protein resulted in comparable egg production from the report of Lopez and Leeson (1993) as compared to a 16% protein control diet. Indeed by the end of the trial, egg numbers produced were similar for both these treatments. However, based on the body weight curves 10% dietary protein was not high enough to maintain body weight at acceptable levels. As has been reported on numerous occasions with egg production type hens low dietary protein levels, while supporting good egg production usually result in smaller egg weights regardless of essential amino acid supplementation. Lopez and Leeson (1993) also demonstrated and increase in hatchability with lower protein diets. It is interesting to note the number of reports that have shown the detrimental effects of high protein intakes for broiler breeders, yet many in the industry still insist on feeding high protein breeder diets.

Pearson and Herron (1981) reported that a ratio of dietary protein to energy of higher than 15g:1M joules resulted in reduced hatchability of breeders. This is of interest as converting their ratio to protein per

100 calories gives a value of 6.28 which is in line with the ratio of protein to energy reported by Spratt and Leeson (1987) to maximise chick hatch weight. There appears to be a lot of evidence to suggest that most broiler breeders are being subjected to excessive intakes of dietary protein. Not only is such a practice detrimental to performance, but it is uneconomical as well as resulting in a greater potential pollution problem with higher levels of nitrogen in the litter from such flocks.

The argument is still heard that "I have to continually increase daily feed allowance in order to maximise egg production", and/or "I had to increase dietary protein levels to increase egg number and size". As has been clearly demonstrated the breeder's main nutrient requirement is for energy and around 80% of this (under good management conditions) goes just to maintain the bird. If the hen's intake of feed is not sufficient to meet her maintenance energy requirement, dietary protein will be preferentially burned for energy purposes.

Hence, a level of dietary protein which should have readily met the hen's total requirement is now deficient with respect to meeting the breeder's requirement for egg mass production. If dietary protein is increased, with such a situation, often an increase in egg mass output is noted. Also if an increase in feed allowance is made a response in performance is usually also noted as more energy is available to meet energy requirements, thus leaving a greater amount of dietary protein to be utilised for egg mass output

What often happens is that dietary protein levels are increased to the point where protein is in excess of requirements and thus nitrogen excretion is increased. It requires a substantial amount of energy to synthesise and excrete uric acid, the nitrogen excretory product of birds. Thus, as dietary protein levels are increased, protein intake increases, uric acid excretion increases and the energy requirement of the bird significantly increases.

With such a situation production responses are often noted with excessive allowances of feed. If a flock is not attaining expected egg numbers and size with a daily feed allowance of 150 to 160 g/b/d, one should take a critical look at possible management factors before changing diet composition or significantly increasing feed allowance. However, it must be remembered that some well managed uniform flocks are reaching production peaks of over 85% and maintaining good production longer than were flocks a few years ago.

Such flocks may require more than the normally recommended level of feed allowance in order to meet their energy requirements. While it is true that as egg mass output increases and a larger percentage of protein intake is partitioned into egg production, one might question whether dietary protein level should be increased. However, with the increased intake of protein as feed allowance is increased, intake of dietary protein is seldom, if ever a problem with broiler breeders.

As mentioned previously there are many good production models available for estimating energy and protein requirements of breeders and in many cases these are considered by nutritionists. The intent of the present article was not to try and compete with the nutritionist regarding diet formulation, but rather to try and point out to producers avenues to pursue and reasons why they may not be achieving optimum performance from their flocks.

It is often stated that seldom is the diet at fault but rather it is the feeding programme or the management conditions under which the diet is being fed which is the problem. This is especially true for broiler breeders where the nutrient intake and requirements of the birds is very much under the control of the flock manger.

Broiler breeders are fed to maximise the production of saleable chicks per bird. Considerations of some of the points raised in the present article might help some producers to increase this number.

The Only Good Broiler Breeder Egg is a Fertilized Egg

Hatchability is around eight percentage points lower than fertility because many chick embryos are usually lost during incubation. For example, even if 93 percent of the eggs laid are fertilized, then under normal incubation conditions only 85 percent of the eggs will hatch. This example illustrates how fertility must be very good to get above average hatchability and hatch bonus pay.

Breeders need to be kept under ideal conditions for maximum life of flock fertility. The chicken's reproductive system is very sensitive to the bird's environment, and under poor conditions the reproductive system will dwindle. For example, the environment can cause a rooster's testes to increase or decrease in size by several hundred fold. But, before we can understand which management factors influence fertility, we must first examine the fascinating process of fertilization in poultry. Fertilization in any animal depends on production of eggs from the female and sperm from the male. A problem with either sperm or egg production can decrease fertility. The rooster's reproductive system is simple when compared to humans or other mammals. The rooster does not have a prostate gland or any of the accessory reproductive glands. Like all other animals, chicken sperm carry the genetic material from the rooster and are produced within the testes. The rooster has two very large testicles within the abdominal cavity on each side of the backbone. After sperm leave the testes, they enter the epididymis, where they gain the ability to swim. Next, the sperm enter the vas deferens, where they are stored until the rooster mates with a hen.

Sperm formation takes about 15 days. The rooster's semen contains around 5 billion sperm per cc, about 40 times as much as that of a human. Once a rooster is mature and if he is maintained properly, he will manufacture about 35,000 sperm every second of his life. However, just like the males of many animal species, the fertilizing potential of roosters varies, even within a flock. For example, some roosters are extremely fertile and create a maximum number of quality sperm; other roosters are subfertile and do not make enough good sperm. This variation in rooster quality is caused by management, environment, nutrition, and genetics. The hen does not produce nearly as many eggs as the rooster produces sperm, but during her 40 weeks of production, the broiler breeder hen lays about 180 eggs. Egg formation requires about 25 hours. Since egg formation requires more than 24 hours, even the best hens cannot lay an egg every day in succession throughout their productive life. As is the case with roosters, some hens are more productive than others, and management has a major impact on variability among hens.

The hen's reproductive system can be divided into two major components: the ovary and the oviduct. The ovary produces the egg yolk. The oviduct adds the white, shell membranes, and shell to the yolk to complete egg formation.

The hen has only one ovary, which is on the left side of her abdomen. The ovary has several thousand ova (egg yolks) in different stages of development and looks like a bunch of grapes. Very immature yolks contain only genetic material from the hen, and as the yolks grow to around 1 mm in diameter, they become white. If the hen is managed properly, many of these developing egg yolks will mature in about 19 days into large, 35 mm, yellow yolks. As the egg yolk develops it will get water, sugars, fats, proteins, vitamins, and minerals from the hen's blood. These are all necessary for the embryo to develop. The egg yolk is surrounded by the perivitelline membrane. This keeps all of these nutrients in a ball-shaped package. One particularly visible region of the perivitelline membrane is the germinal disc. The germinal disc is a small white dot about half the size of a pencil eraser on the surface of the yellow egg yolk. Fertilization takes place here, and embryonic development begins.

When the egg yolk is mature, it leaves the ovary, and within 20 minutes it is captured by the infundibulum, the first part of the oviduct. Here fertilization takes place. Following mating, sperm enter the hen's oviduct and are stored within sperm storage glands. Only sperm that can swim will enter these sperm storage sites. These glands can store more than half a million sperm. Sperm can remain alive in these glands and fertilize eggs for up to 3 weeks.

A hen will have maximum fertility for only about 3 to 4 days after one mating. For this reason, the male-to-female ratio in a flock must be enough to ensure mating of every hen every 3 days or so. Sperm do not break through the eggshell. Instead they travel up the oviduct to the infundibulum to join with the egg yolk. The sperm bind to the perivitelline membrane and make a hole as they enter the egg. Hundreds of sperm may enter the yolk. As a matter of fact, the more sperm that enter the yolk, the more likely the egg will be fertilized. Around 30 sperm must enter the egg near the germinal disc to insure a 95 percent chance of fertilization. While it is true that only one sperm is necessary to fertilize an egg, the probability of an egg's being fertilized by only one sperm's reaching and penetrating it is very low.

After about 15 minutes, the yolk leaves the infundibulum (fertilized or not) and receives the egg white, shell membranes, and shell over the next several hours from the magnum, isthmus, and uterus sections of the oviduct. When the hen lays a fertilized egg, the chick embryo has already developed for about 25 hours into approximately 20,000 embryonic cells and is a live, breathing organism. If this fertilized egg is handled properly before and during incubation, a healthy baby chick is the result. This publication is a joint effort of the Mississippi State University Extension Service and the Mississippi Agricultural and Forestry Experiment Station.

Weighing Broiler Breeder Females Post-Feeding

Obtaining accurate body weights is a critical part of the process of rearing replacement broiler breeder pullets and managing breeder hens and males. From the first few weeks of age in the pullet house, all feed allocations are determined by the bird's weekly weight gains. Obtaining accurate body weights is very important to maintaining uniformity, body conformation and the overall development of pullets and young cockerels. Research has shown that accurately and uniformly controlling body weight of both replacement breeders and breeders in the hen house will result in improved performance parameters.

In the United States, the majority of poultry integrators rear pullets on some version of a skip-a-day feed programme in order to control body weight among all the birds in a house. Under our current housing conditions, skip-a-day feed programmes are the best way to uniformly distribute feed to all birds simultaneously in an effort to maintain body weight uniformity. However, the presence of feed in the crop or digestive tract will inflate the actual body weight of the birds and skew feed allotments.

Therefore, replacement breeders are typically weighed on off-feed days to normalise the data and not confound body weights with either the presence or absence of feed in the crop or digestive tract. This allows for body weight measurements to be consistent from week to week without regard for feed clean up time and the presence or absence of feed in the crop. Therefore, each week pullets and cockerels are weighed with an empty crop and digestive tract. This process continues until birds are moved to the hen house and feeding begins on an everyday basis.

These weights are considered to be 'empty' weights. In the hen house, most commercial producers move from a skip-a-day to an everyday feed programme as hens are brought into production. Feed is often provided daily in the early morning hours shortly after the lights are turned on. While feeding hens everyday in the hen house has proven to be an effective management tool, birds cannot be weighed on 'off feed' days.

This has led to the concern over whether hen weights are truly reflective of the actually body weight and mass. Consequently, current industry recommendations are designed to address this issue and suggest producers weigh breeders late in the afternoon hours to obtain the 'empty' weights. This allows any feed consumed to have time to pass through the birds digestive system and therefore create an 'empty' weight situation for weighing purposes. In breeders this can be further complicated by the fact that the majority of egg production occurs in the morning hours following feed cleanup which would result in additionally body weight loss. To address this issue, a research project was designed to weigh breeders at various intervals during the day to determine the best time to weigh birds to most accurately reflect actual body weight gains.

When to Weigh Breeders

Birds used in this study were housed at the University of Arkansas Broiler Breeder Research Farm. A single pen of breeders containing 71 hens was used for this study and during each weigh period, all hens were corralled in a catch pen with each hen weighed individually so that no sampling error could affect the results. All hens were weighed prior to daily feeding and again at feed clean-up time. Additional bird weights were obtained at 2, 4, 6, 8 and 10 hours following feed clean-up. This process took place on the same birds at 24, 28, 34 and 41 weeks of age. These age periods represented pre-laying, pre-peak, peak and post-peak in production stages of life.

It was previously believed that hens would lose body weight throughout the day to approach the 'empty' weights found prior to feeding. However, these data make it apparent that the passing of feed and the consumption of water appear to offset each other and allow the hen to maintain a near constant body weight through 10 hours following feed clean-up. Body weights obtained prior to feeding would be the only weights that could be considered 'empty' weights as they were obtained immediately after lights came on in the morning and are a reflection of body weight loss due to feed and water passage occurring during the dark hours. These results would allow breeder service technicians to weigh breeders in the hen house at any time following feed cleanup and that the data would be consistent with body weights obtained at any time throughout the day. These data will allow technicians to be more productive in a given day in regards to scheduling weighing of breeders in the hen house.

How to Weigh Birds

When weighing birds, it is often recommended to weigh all birds caught in a catch pen and not weigh a specific number of birds to meet a given criterion. This has been the recommendation for broilers in research trials but has not been evaluated in replacement pullets and breeders. As part of this project, body weights were recorded for each hen in the order they were caught in the catch pen. For each age group and for each time interval previously mentioned, this resulted in 40 incidences of weighing all birds in a catch pen. For instance, if 60 birds are caught in a catch pen and only the first 50 are weighed because that meets the minimum number needed, then the body weight recorded would not be reflective of the actual weight of the birds caught or the birds in the flock. If this occurs with pullets and feed allotments are determined based upon these body weights, then inaccurate feed allotments could be provided and less control over flock body weight would be the result.

Summary

1. When weighing broiler breeders in the hen house, accurate and consistent body weights can be achieved by weighing birds at any time after feed clean-up. There is no advantage to waiting for feed passage in an attempt to obtain 'empty' weights in breeders during the afternoon hours.
2. When weighing birds caught in catch pens, it is important to weigh all birds caught in the pen and not stop at a predetermined number of birds. The last birds caught will be the smallest birds and need to be included in the final group weight to most accurately determine the average body weight of the birds in a flock.

Health Implications for Higher Density Broiler Production

Such actions are making us acutely aware of the "law of diminishing returns." We are now faced with a more accurate evaluation of the application of the "law" to the broiler chicken of today, the facility in which the chicken is grown, the feed the chicken is fed and the grower who manages it all.

The broiler chicken of today is truly remarkable. The superior genetics has resulted in increased body weight gains, increased yields, and increased efficiency of feed utilisation. In Western Canada, managing the rapid growth rate has become our number one priority. To accomplish this task, the grow facilities are usually equipped with computers to consistently control the bird's environment. Feed formulations have been improved. Water and feed delivery systems are being improved. All of this investment in equipment and time will ensure the bird reaches its full potential. But, will this still be the case with increased bird densities?

The health implications for higher density broiler production are significant and must be taken into account. Bird welfare, in my opinion, must also be considered. Bird welfare, is not going to be, it is, extremely important in terms of meeting the consumer demands of today and tomorrow. What happens when we add more kilograms of live weight to a given, fixed amount of floor space? As we examine the possibilities, we should always evaluate our FLAWS. With increased density, feed and water will become more difficult for each bird to access. This will lead to reduced performance on each normal bird. Furthermore, birds which may only have a marginal disability will become less able to compete as the stocking density rises.

Poorer litter conditions associated with higher moisture content occur with higher stocking densities. Higher litter moisture will "increase the wave of insult" by increasing the concentration of pathogenic organisms.

Reduced access to feed containing anticoccidials/antibacterials together with increased exposure to oocysts/bacteria may result in clinical coccidiosis and/or bacterial infections. Poor litter conditions can also exacerbate leg disorders experienced by chickens. Poorer litter conditions, reduced access to feed, and an increased demand for fresh air may result in an increase incidence of pulmonary/ cardiovascular disease. Presently, we are faced with an ever increasing incidence of pulmonary hypertension (ascites). By increasing the stocking density we will be increasing the demand for vital oxygen. Increasing the demand for oxygen will add more pressure to the bird's pulmonary and cardiovascular systems.

Increasing stocking density will increase the likelihood of a bird being scratched. This coupled with an increase in litter moisture will set up the possibility for an increase in the incidence of type II cellulitis. There are many "roads that lead to the cities of ascites and cellulitis" and increasing the stocking densities will only make "these roads easier to travel." The ability to vaccinate our birds via the drinking water will be compromised by increasing the stocking density. Poorly vaccinated flocks are more prone to vaccine "rolling" reactions and are more prone to disease. Increased stocking densities will increase stress. Increased stress will manifest itself in many ways, most commonly as a reduction in overall performance. Increased stress will result in an increased susceptibility to the common broiler diseases in a given geographical area and may open the door for new and re-emerging diseases.

If we increase the stocking density of our broiler chickens beyond our capabilities as broiler farm managers, then we will suffer losses. If we increase our capabilities as broiler farm mangers, then we will be able to grow more chickens in less space. Let us first increase our capabilities, then add more chickens. As a poultry veterinarian today, the majority of my work has revolved around solving "problems" that involve more than one etiology or cause. It clearly involves concentrating on prevention. "Magic bullets" do not exist.

The diseases that we are faced with "preventing" have multiple and convoluted causes and involve the understanding of all disciplines of broiler production.

Our broiler chickens have served us extremely well in the past and by all present indications our broiler chickens today are performing even better. More commitment will be required by the entire broiler production team. This team will be required to operate and work together in an orchestral fashion in order to achieve success.

6

Organic Chicken Farming

Organic chicken farming has become more prevalent as the organic- and local-foods movements gain momentum. Farmers and consumers should be aware of the pluses and minuses involved in raising organic chickens. The term "organic" and the organic movement have been around for some time, but in recent years, the term's scope has broadened. Many farmers tending livestock, such as chickens, have worked hard to limit the amount of hormones, drugs and chemicals going into their animals' systems, but organic is not always synonymous with the best practices.

Significance

A farmer is raising organic chickens if he is supplying his livestock with feed that is blended from organically grown products. However, just because a chicken is fed organically does not mean it is receiving proper nutrition. Many farmers use corn to help increase the weight of chickens meant for meat. Organic farmers choosing to feed corn will use organic corn, but chickens should not be solely or heavily fed corn on a steady basis; they need a mix of grains, greens and protein. In addition, raising chickens organically should include preventative health measures, and the treatment of diseased animals should be approached holistically and non-chemically. Although many studies have been done on the benefits of eating organically, little has been proven absolutely. Poultry and eggs that do not contain hormones or other chemical products are believed to be healthier because they do not pass along those items or their byproducts to the consumer. There is little statistical significance that other food-borne illnesses, like salmonella, are decreased or increased by the farming method employed.

Misconceptions

Chickens and eggs marketed as organic do not automatically come from animals that have been raised cage-free or free-range. Because the organic market is becoming big business, some organic farms employ conventional tactics. Large organic farms may keep their chickens using the same confinement methods as conventional farms, which may cage chickens for their entire lives, so that they have little or no access to the outdoors. In contrast, many farms that feed their animals non-organic feeds and/or use hormones and drugs to treat their animals may allow those chickens to live cage-free.

Potential

Using organic-farming techniques to raise chickens and eggs will, if nothing else, improve the health of the animals and the ecology of the farm. Studies do indicate that eating organically helps significantly reduce the amount of pesticides, hormones and unwanted chemicals in our diets. How to Organic free-range chickens are happy chickens. Farmed chickens are all to often kept in crowded quarters, rarely see daylight and are fed grain treated with antibiotics. Purchase your chicks from local farmers who also raise free-range chickens to insure your flock is healthy from the beginning. You can have the benefits of free-range chicken and eggs by raising them yourself.

How to Raise Organic Free Range Chickens

Feed a diet of whole organic grains, or premixed organic feed. It is important that the feed is truly organic and does not have any added antibiotics. You can purchase ready-made organic feed or look for recipes online to mix your own organic feed.

Some rare breeds of chickens are in need of conservation. Jupiterimages/Photos.com/Getty Images Allow your chickens free range. The chickens will rely less on grains if they are encouraged to forage for their own food on grass and are allowed to roam a wide area on a daily basis. In the warm months, free-range chickens require less feed reducing the expense of raising them and at the same time ensuring healthier chicken. Free-range chicken should spend a minimum of two hours outside each day, but more time outdoors is greatly encouraged. Two square feet of outdoor ground per chicken is a minimum amount of space that chickens must be provided to be called "free range." But, that is a very small amount of space and ideally a chicken should be given much more. Ten square feet per chicken is a much more reasonable amount of space. Chickens that are not crowded peck at each other less and are less prone to mites and disease. Some people let their chickens roam freely, but they are at the mercy of predators and they must be trained to return to the coop at night. Free range chickens may be allowed to roam a fenced-in area if they are given adequate space. Some people build pens that can be easily moved to allow the chickens to range new grass periodically.

Supplement your chicken's feed with stale bread and table scraps. Add crushed garlic or unfiltered apple cider vinegar in their water periodically. It is a natural way to prevent infection and disease. Do not give this to them in a galvanized water container. The chickens will also need grit, sand or small gravel.

The Benefits of Organic, Free-Range Chicken

Organic, free-range chickens are often thought to be more healthy for consumption than chickens from other farms. Raised without using any hormones or antibiotics, the chickens have a more natural existence. The feed they eat is all organic and grown without pesticides. After the chickens are killed, no additives are introduced into the meat.

Stress-Free Chickens

Because the chickens are living organically and free-range, they aren't as stressed out as chickens raised in cages. The chickens have more space instead of being raised in tight crowded quarters next to thousands of other chickens. Disease is less apt to spread when chickens are raised free-range and their lives are less stressful. Because they aren't as stressed, they live healthier lives and are healthier for people to eat.

No Antibiotics

Most chickens raised in the meat production industry are given antibiotics. The antibiotics given to chickens are then ingested by consumers when the chickens are cooked and eaten. These antibiotics could have unintended effects on the people who eat them. The antibiotics could make bacteria and infections more resistant to treatments and create stronger strains of bacteria. Organic, free-range chickens aren't given antibiotics, and are also free of hormones. When chickens are able to grow healthy and naturally, their meat is in better condition and has more flavour than meat from conventionally grown chickens. The chickens are able to get more exercise, which is healthy for their bodies and improves the quality of their meat. Also, they receive better nutrition, which also contributes to the flavour of the meat. The end result is chicken meat that is juicy, moist and rich in flavour.

It Shouldn't Be This Much Fun To Keep Happy Healthy Chickens In Your Own Backyard... But It Is!"

If you're interested in keeping chickens at home, or you already are, then Chicken Keeping Secrets is here to provide you with litterally everything you need to know. Over the past 3 years, we've helped 102 s of thousands of home chicken keepers from around the world with all of their chicken keeping needs, and now we're here ready and willing to help you too...

With rising food prices, uncertainty about chemicals & additives in our food, and the way 95% of chickens are inhumanly treated with intensive "cram 'em in cages" farming... it's no wonder. Intensive chicken farming is just plain wrong, but I know you already feel the same way...And here's a little bit of info that might gross you out...

Did You Know...

The average egg purchased from your local supermarket or convenience store, whether it be from a barn raised or free range chicken, is generally more than 45 days old before it ends up on your table? Try eating an egg that's less than 24 hours old – you'll soon taste the difference...

When you keep chickens at home, you never have to worry about old stale eggs that have lost most of their nutritional value, and more importantly, you'll never be supporting the intensive chicken farming industry again. You just can't beat that. However, while most people have good intentions, they also have a tendency to rush into chicken keeping. They think that keeping chickens in their own backyards or on a small rural plot is like, well, chicken feed. Unfortunately keeping chickens in your own backyard is not as easy as keeping the shrink wrapped variety that might often be accustomed to visiting your freezer.

However... the good news is that IT CAN BE that easy when you're armed with the right knowledge if there was such a thing as a doctorate in keeping chickens, that even a country yokle from the backwaters of timbucktoo could follow with ease – Chicken Keeping Secrets would be it...Private access to the "Chicken Keeping Secrets" membership website is like having your own personal library of proven chicken keeping information & advice, right at your finger tips, that you can access from the comfort of your own home 24 hours a day, 7 days week. Whether you need instant help with chicken housing information, health related issues, chicken feeding & nutrition problems – in fact any chicken keeping issue at all – we've got you covered. Chicken Keeping Secrets can help you from A-Z... whether you're a complete beginner and don't know where to start, or you're a seasoned chicken keeping professional and just want practical "how to" advice on tap.

Introducing "Chicken Keeping Secrets" brand new, official guide to keeping chickens at home which you can access and download in just a couple of minutes...The Chicken Keeping Secrets guide was not only thorough and to the point, but also extremely informative. I particularly liked the chapter on chicken health. The way you've covered all of those elements in detail is certainly going to be of help in the long term. And those bonuses, awesome. I would happily recommend membership to anyone thinking about keeping chickens – it's fantastic and I'm happy I signed up for sure." ~ Louisa Gagliardi, Queensland, Australia

In this concise, easy to read, step by step guide – that you can download instantly in a couple of minutes from now - you'll discover... The 10 most commonly kept breeds of chicken including; their suitability for egg laying and/or meat production, their basic requirements and adaptability to your specific climate, and perhaps most importantly if you have children – their different temperaments and personalities. A brief history of chicken keeping and how to determine whether keeping chickens in your own backyard is really the right option for you. A complete run down on what chickens need to really thrive, the costs involved (they're much lower than you might think), and how much time you're really going to need to dedicate to the new additions to your family...

We've had chickens since I was about 5 years old (we've got a soft spot for silkies) so I wasn't sure if I really needed to purchase a membership to Chicken Keeping Secrets in addition to getting your free newsletter. However, sometimes it's nice to be able to grab a quick second opinion when unusual issues arise. I think this is great value for money, and it's good to know I can come in here at anytime and search around for your professional opinion on things. You've done a great job here" ~ Neil Patchen, New York, USA. A detailed discussion on chicken coops including size and space required, accessibility issues, ventilation and insulation, and the placement of your coop. You'll find out how much space you really need and the answer might just surprise you. Should you buy a ready made chicken coop? Should you build your own from scratch or knock up a kit set coop? Does a chicken really care if it lives in an old dog kennel? We look at all these answers and more to help you make an informed decision that will keep your chickens, your wallet... and your "significant other" happy. Choosing the right bedding material to use on the floor of your chicken coop is extremely important. Get it wrong and the health of you chickens may suffer. Get it right and you'll reduce smelly odors, maximise hygiene, and minimise cleaning time and requirements. We discuss various bedding material options here and look specifically at those that you can then use as high yielding organic compost for your garden. Are you going to let your chickens out every day to forage around your yard so they can scratch for bugs, grubs, and insects becoming a natural form of pest control? Or should you keep them secure in their own chicken run? If you do let them out how are you going to control them? You'll find out everything you need to know here.

Your new hens are going to need a place to nest inside their home so they can reward you with fresh eggs daily. We look at different types of nesting boxes, their size requirements, positioning, and the ideal number to have to maximise egg production without taking up unnecessary space in the coop, or increasing your hobbies expenditure. Did you know that chickens don't actually sleep inside their nest boxes? At night they actually roost on perches so you'll need to ensure their new home has a suitable roosting place. We discuss different perching options, including thickness requirements, positioning to reduce fighting for the "prime roost", the correct height that they should be placed, general spacing requirements, and common materials.

Having the right tools and equipment for your new flock is a must to ensure harmony amongst "the pecking order". You'll learn about suitable food and water dispensers, how many you should have depending on how many birds you keep, where they should be placed, and whether you should go for the cheaper plastic varieties or the robust but more expensive options.

Natural and artificial lighting inside your chicken's home is one of the most important factors that needs to be considered for the health of your birds. Not enough light and egg production will be reduced. Too much light and you'll wear your hen's health down. Here we look at the various lighting options that are available and the ideal lighting balance required to keep things in check...

We take a detailed look at the chicken's digestive system, how it functions, and what's natural so you can keep a close eye on things to ensure everything's normal. How much should you feed your chickens? When should you feed them? Should you use a commercially brought feed or a home made feed? Here you'll receive a detailed education on the proper nutritional requirements for maintaining a healthy chicken based on differing needs during its lifetime. If you want to make your own chicken feed we provide you with 3 easy homemade chicken feed recipes. Now you'll know exactly what you're putting into your chickens and ultimately what's going to end up inside you when you consume their eggs.

If you're going to supplement your chickens feed with your own household scraps you'll learn what you can and can't feed your chickens. Some household scraps can become real treats for your flock; others can be detrimental and even fatal to their health and should never be given to them. We take a detailed look at chicken pellets, chicken mash, cereal feeds, greens, calcium and other vitamin and mineral supplements, chicken grit, "scratch" and much more. Knowing how your chickens should and should not behave in their regular daily routines is fundamental to ensuring everything is fine. You'll learn about your chickens common behavioural habits such as broodiness, preening, roosting, and more.

You'll learn about the "pecking order" and what to do when you see one of your hens being picked on by other birds in your flock. The "pecking order" is completely natural in chickens but often it can get out of control. You'll learn exactly what to do if one of your hens becomes too aggressive and the bird on the receiving end becomes injured or hurt. You'll also learn about less common behavioural habits such as feather pecking, egg eating, cannibalism and more. Nip these last ones in the bud before they get out of control so one of your girls doesn't end up getting seriously hurt, or even worse, potentially killed. We take a look at 3 of the most common minor aliments that can impact on the health of your chickens, and then a whooping 25 serious aliments that might impact them at some point in the future. Each includes a full description of the aliment and appropriate suggestions on treatment and the course of action you'll need to take to keep your chickens in tip top condition.

We take a look at 8 of the more common types of parasite that can take a fancy to our feathered friends, how to minimise their occurrence in the first place and what to do if they've already arrived and are irritating your chickens.

Thorough cleaning and hygiene will dramatically reduce any associated health problems and is therefore critically important to the overall health of your flock. We look at how to properly clean out your chicken coop, and what types of cleaning martial you should use to ensure the safety of your birds – whilst at the same time maintaining the cleanest and most hygienic housing possible the above is just a small fraction of what you're about to learn. The "tip of the chicken feather" so to speak. Our exclusive guide to chickens – "Keeping Happy, Healthy, Egg Laying Chickens In Your Own Backyard" contains over 97 pages of pure, no fluff, particial chicken keeping advice. This is your complete guide to keeping chickens in your own backyard.

However, this guide is NOT available in any stores – nor on any other websites. We're only making this exclusive manual available to the valued visitors and newsletter readers of the "Chicken Keeping Secrets" website. BUT the good news is... THAT'S YOU... and because we've produced the manual in digital format, it's available for you to download to your computer and read right now, even if it's 2am in the morning. What's even better is that your investment in this manual (and access to a very exclusive private membership area) is just a small, one time payment of only your purchase is protected by my personal. If for any reason whatsoever you're not totally convinced that this is the most informative guide to chicken keeping you've ever read, simply drop me an email within the next 60 days & I'll refund 100% of your purchase price – no questions asked;

You've got absolutely nothing to loose with a guarantee like that, and if the health & well being of your feathered friends is important to you, then you realisese that $27 is just a small drop in the bucket compared to sharing a long healthy & productive life with them.

Heck, you've probably spent more on a decent feed at K.F.C;-0You will be able to download "Keeping Happy, Healthy, Egg Laying Chickens In Your Own Backyard" immediately after payment has been made and you've completed registration for the private members area (a 30 second process). The download itself is a PDF document which can be viewed on either a PC or MAC – if you don't know what a PDF document is dont worry, we'll provide you with complete help.the "Chicken Keeping Secrets" Guide To "Keeping Happy, Healthy, Egg Laying Chickens In Your Own Backyard" certainly stands on it's own two feet, but as we mentioned before, membership to "Chicken Keeping Secrets" is more than just a simple "guide to keeping chickens". Membership is like having your own personal library of proven chicken keeping information & advice, right at your finger tips, that you can access from the comfort of your own home 24 hours a day, 7 days week. For a limited time only, we're also going to include several fast action bonuses inside a very exclusive private membership area when you order now... We've sold hundreds of copies of this classic collection as a stand alone product, at the same small price you're going to invest today for complete membership.

But you wont have to pay anything extra, the entire collection is available in our exclusive customer only area for free, and like our own exclusive guide to keeping chickens, all volumes are available for immediate download...

Hi Duncan, I wanted to thank you for the collection... initially I had my reservations with the books being written quite some time ago, but you were right. They are excellent reference material and these guys certainly know what they're talking about... I was so impressed with "The Practical Poultry Keeper" I actually took it down to the local printers and had it spiral bound. It's now sitting on the shelf in the chicken shed to reference whenever I need it. Thanks again & I look forward to next issue of newsletter" ~ Mary Beaven, Austin Texas. This massive 344 page publication was, and still is, considered one of the most in-depth and comprehensive guides to successful backyard chicken & poultry keeping.. If you're interested in really maximising the egg production of your backyard chicken flock then this guide is exactly what you've been looking for. Not only will you never be short on eggs again, but with 200 eggs per year, per hen, you'll have enough left over to sell to your neighbours and friends and make a nice tidy profit. There is nothing quite like a hobby that not only pays for itself... but actually puts money back in YOUR pocket.

This publication will teach you how to make and use time saving chicken keeping & management devices to ensure you're chickens require as little of your time as possible. The less time you spend looking after your chickens, the more time you'll have to enjoy them! This 130 page publication was considered the authoritive guide to practical chicken coops in it's day. With over 100 detailed illustrations, if you're looking to build your own chicken coop, this book is an essential reference. This detailed 116 page publication will teach you everything you need to know about chicken coops & hen houses... "The Poultry Doctor" contains absolutely everything you need to know about the health of your chickens...Making A Poultry House is yet another short and concise guide to making your very own chicken coop or hen house he "Masters Of Chicken Keeping Collection" is a must have resource for anyone interested in keeping chickens. Written by the masters of the day – when just about everyone kept a few chickens in their own backyards – these guys really know their stuff. Thanks. I have downloaded volume one and must do the others. It is such a great read I haven't taken the time to download the rest. I agree with the comments. I too thought it was a bit old! But what a great read and with the environment so crazy it's good to read/learn about the practical 'how to do' (techniques) especially if one doesn't want to buy feed that may contain things one doesn't want... I was surprised to learn that we really didn't know all that much about chickens – we have had some before. Now rather than rushing out and buying some we are rethinking the how's and whys of the project. This time I am sure we will do a much better job! Thank you to you" ~ Dale Mattock, Canada.

Again, this entire collection can be downloaded immediately after ordering our exclusive guide to keeping chickens at home and is included at no extra cost...

Intensive Chicken Farming

In egg-producing farms, birds are typically housed in rows of battery cages. Environmental conditions are automatically controlled, including light duration, which mimics summer daylength. This stimulates the birds to continue to lay eggs all year round. Normally, significant egg production only occurs in the warmer months. Critics argue that year-round egg production stresses the birds more than normal seasonal production. Meat chickens, commonly called broilers, are floor-raised on litter such as wood shavings or rice hulls, indoors in climate-controlled housing. Poultry producers routinely use nationally-approved medications, such as antibiotics, in feed or drinking water, to treat disease or to prevent disease outbreaks arising from overcrowded or unsanitary conditions. In the U.S., the national organization overseeing chicken production is the Food and Drug Administration (F.D.A.). Some F.D.A.-approved medications are also approved for improved feed utilization. In the U.S., federal law prohibits the use of hormones or steroids in poultry production.

In egg-producing farms, cages allow for more birds per unit area, and this allows for greater productivity and lower space and food costs, with more efforts put into egg-laying. In the U.S., for example, the current recommendation by the United Egg Producers is 67 to 86 in^2 (430 to 560 cm^2) per bird, which is about 9 inches by 9 inches. Modern poultry farming is very efficient and allows meat and eggs to be available to the consumer in all seasons at a lower cost than free range production, and the poultry have no exposure to predators. The cage environment of egg producing does not permit birds to roam. The closeness of chickens to one another frequently causes cannibalism. Cannibalism is controlled by de-beaking (removing a portion of the bird's beak with a hot blade so the bird cannot effectively peck). Another condition that can occur in prolific egg laying breeds is osteoporosis. This is caused from year-round rather than seasonal egg production, and results in chickens whose legs cannot support them and so can no longer walk. During egg production, large amounts of calcium are transferred from bones to create eggshell. Although dietary calcium levels are adequate, absorption of dietary calcium is not always sufficient, given the intensity of production, to fully replenish bone calcium.

Under intensive farming methods, a meat chicken will live less than six weeks before slaughter. This is half the time it would take traditionally. This compares with free-range chickens which will usually be slaughtered at 8 weeks, and organic ones at around 12 weeks. In intensive broiler sheds, the air can become highly polluted with ammonia from the droppings. This can damage the chickens' eyes and respiratory systems and can cause painful burns on their legs (called hock burns) and feet. Chickens bred for fast growth have a high rate of leg deformities because they cannot support their increased body weight. Because they cannot move easily, the chickens are not able to adjust their environment to avoid heat, cold or dirt as they would in natural conditions. The added weight and overcrowding also puts a strain on their hearts and lungs. In the U.K., up to 19 million chickens die in their sheds from heart failure each year.

Issues with Poultry Farming

Humane Treatment

Animal welfare groups have frequently criticized the poultry industry for engaging in practices which they believe to be inhumane. Many animal rights advocates object to killing chickens for food, the "factory farm conditions" under which they are raised, methods of transport, and slaughter. Compassion Over Killing and other groups have repeatedly conducted undercover investigations at chicken farms and slaughterhouses which they allege confirm their claims of cruelty. Conditions in intensive chicken farms may be unsanitary, allowing the proliferation of diseases such as salmonella and E coli. Chickens may be raised in total darkness; hens are most often kept in crowded wire battery cages with space less than that of a sheet of paper per hen, as opposed to cage-free or free range. Rough handling and crowded transport during various weather conditions and the failure of existing stunning systems to render the birds unconscious before slaughter have also been cited as welfare concerns. Another animal welfare concern is the use of selective breeding to create heavy, large-breasted birds, which can lead to crippling leg disorders and heart failure for some of the birds. Concerns have been raised that companies growing single varieties of birds for eggs or meat are increasing their susceptibility to disease. A common practice among hatcheries is the mass-slaughter of newly-born male chicks of egg laying breeds, since they don't lay eggs, and do not grow fast enough to be profitable for meat. Once separated from the females, the male chicks are dropped into grinding machines.

Debeaking

Laying hens are routinely de-beaked when young to prevent fighting. Because beaks are sensitive, the usual practice of trimming them without anaesthesia is considered inhumane by some. De-beaked chickens will peck much less than chickens with beaks, which animal behaviourist Temple Grandin attributes to guarding against pain. The United Egg Producers says that de-beaking is not painful. It is also argued that the procedure causes life-long chronic pain and discomfort and decreased ability to eat or drink. De-beaking is described as being as painful as for a human's fingertips, rather than nails, being cut off.

Facts about the Poultry Industry

The average consumer may not be aware of the suffering of billions of birds raised for meat and egg production in the United States each year. The U.S. Department of Agriculture (USDA)'s National Statistics Service reported that 7.07 billion "broiler" chickens, 67 billion "egg" chickens, and 321 million turkeys were killed in 1998 for food. In addition, millions of birds die as a result of disease, injury, and during transportation.

Egg Production

Egg-laying hens in the United States number more than 459 million. Of these millions of birds, 97% are confined to "battery" cages, tiny cages roughly 16 by 18 inches wide. Five or 6 birds are crammed into each cage, and the cages are stacked in tall tiers. As many as 50,000 to 125,000 battery hens, in sheds with minimal light, strain to produce 250 eggs per year, ten times the number of eggs they would produce in the wild. Battery cage confinement does not allow birds to turn around or take part in any other natural behaviour, such as preening, dust bathing, and foraging for food. Prolonged forced confinement causes unnatural behaviours such as cannibalism and increases the incidence of disease and injury. Laying hens are also forced to live in a polluted environment due to toxic feed ingredients, accumulated feces, and excretory ammonia fumes. A successful battery system relies heavily on antibiotics that are routinely administered to laying hens to decrease the incidence of disease among these immune-repressed birds.

Battery hens often die in their cages as the result of disease or injury. Those who survive but stop producing adequately are considered "spent" hens and are sent to slaughter to be used for human and animal food. Male chicks are of no value to egg producers. Each year more than 200 million male chicks are killed or left to die after hatching. Egg-producing birds that are not confined to battery cages seldom fare much better. Eggs labelled "Cage Free" or "Free Range" simply mean that the birds are not confined to battery cages, not necessarily that the hens are allowed a more natural existence. Neither guarantees that they have adequate space to move around, or that they are allowed outdoors to roam about and forage or dust bathe.

Forced Molting

Molting is the natural process of shedding old feathers and the growth of new feathers. Molting initiates a new egg-laying cycle. The natural molting process takes about four months to complete. However, on factory farms, poultry producers induce starvation to control egg production in laying hens (eggs for human consumption) and breeding hens (eggs that hatch into birds used for meat or egg production) to reduce the molting period to one to two months. Performed to increase farm profits, this "forced molting" is extremely stressful to hens. Forced molting methods include food and water deprivation, medications, and simulated light and dark cycles. A 1992 *Poultry Science* report found that forced molting in combination with a Salmonella infection created an actual disease state in tested hens. Salmonella infection can be passed on to consumers through egg consumption.

Debeaking and Toe Clipping

Debeaking is a painful procedure whereby the bird's sensitive beak is sliced off with a hot blade. Poultry meat and egg producers that use battery cages and crowded floor systems remove one-half to two-thirds of the birds' beaks to discourage cannibalistic pecking, a behaviour that occurs when birds are kept in close confinement with no regard for their natural behaviours. Behavioural studies indicate that debeaked birds are often unable to eat, drink, and preen properly. They also exhibit behaviours associated with chronic pain and depression. Toe-clipping is the amputation of a bird's toes just behind the claw. This painful procedure is performed to reduce claw-related injuries on factory farms.

Meat Production

Genetic engineering of broiler chickens and turkeys often results in a bird too heavy to stand or walk. They suffer from pain in their legs and sores on their feet that are induced by their extreme, unnatural size. Kept in polluted dark sheds with as many as 25,000 birds per shed, these birds suffer many of the same ailments as battery hens, such as being debeaked and being forced to live in a toxic environment. Thousands of these birds never make it to slaughter — they will die while still on the farm from injuries, disease, or their inability to reach food and water.

Transport and Slaughter

Millions of birds die during the loading of trucks and while en route to slaughter. These sensitive birds, often in very poor physical condition, are grabbed by their legs and thrown into densely packed cages to be transported by truck to slaughterhouses that are sometimes hundreds of miles away. Many die from shock, injury, and suffocation in the process. The U.S. Federal Humane Slaughter Act does not apply to poultry, meaning that there is no federal law that requires birds to be stunned prior to slaughter. This allows for diversity in commercial poultry slaughter approaches and stunning equipment. When slaughterhouses do use stunning equipment, lack of regulation often results in birds allowed to raise their heads prior to reaching the water bath stunner and therefore not adequately stunned. Problems also exist in neck-cutting equipment, which may result in prolonged and extreme pain caused by necks improperly cut during the killing process.

The Effects of Forced Molting Nutritional Methods on Semi-heavy Layer Performance During Such Period

Molting is characterised by a period of time in which hens experience a reduced or even halted egg production, aiming to renew their feathers and restore their reproductive tracts. This phenomenon typically occurs during decreasing photoperiod times (winter), and it can last for nearly 4 months (Silva, 2000). Nevertheless, this does not assure that birds will continue laying eggs during a second cycle.

The technique of forcing layers to molt in an attempt to optimise egg production for one additional cycle, was first used in the US in the 1960s, but currently several other methods are being evaluated. In the practice, the most popular methods include several different management practices. The feed restriction method is most popular due to its low cost and high efficiency (Rodrigues *et al.*, 1995). In Brazil, the long-fasting technique (i.e., California method) is commonly used, but it has been questioned lately since it badly affects animal welfare. This is why the search is ongoing for alternative methods in agreement with both market perspectives and farm convenience.

In accordance with Lee (1982), when forced molting is practiced properly, it results in increased laying rates and improved internal/external egg quality during the second cycle, as compared with the late stages of the first laying cycle. Among the major consequences of molting, despite of the use of alternative methods, body weight losses occur thus interfering with the late stages of the first laying cycle. In addition, high mortality rates occur. The objective of this research was to study the effects of applying various forced molting methods on the productive performance of commercial layers.

Materials and Methods

The study was carried out in the experimental farm, Visconde de Graça Agritechnical Cluster (CAVG/UFPel), Brazil. Four hundred and forty eight (448) 100-week-old Hisex Brown hens were divided into two 32-cage batteries within a blackout house. The lighting program included 8 hours of light per day. The experimental design was completely at random, and the experimental unit was one 7-bird cage. Each treatment included 16 repetitions. The experimental period lasted for 14 days. Birds were subjected to 4 treatments: T1, 10-day fasting (control group) plus 4 days with ground corn; T2, soy hulls + vitamin premix; T3, wheat bran + vitamin premix; T4, high tannin sorghum + vitamin premix. In the fed groups, feed was given daily at the rate of 100 g per bird. Water was given *ad libitum*. In T2, T3 and T4 the feeds were offered for 14 days on a row. The variables evaluated in this period included feed intake (FI) recorded daily, live weight (LW), weight loss (WL) in the 10-day period, egg (ED), livability (LIV) and mortality (MORT).

Trace Mineral Nutrition in Poultry and Swine

Trace minerals such as zinc, copper, and manganese are crucial for a wide variety of physiological processes in all animals. The biological importance of zinc is underscored by the understanding that zinc is a required co-factor for several hundred enzymes, representing all six enzyme classes, and a wide variety of transcription factors (Vallee and Falchuk, 1993; Underwood and Suttle, 1999). Indeed, proteins with zinc-binding domains are estimated to represent approximately 10% of the human proteome, and the same would be expected to be true in production animals as well (Andreini et al., 2006). As such, zinc plays essential roles in a wide array of processes including cell proliferation and animal growth, immune development and response, reproduction, gene regulation, and defence against oxidative stress and damage (Shankar and Prasad, 1998; Underwood and Suttle, 1999; Fraker et al., 2000; Blanchard et al., 2001; Ibs and Rink, 2003; Song et al., 2009). Zinc's role in gene regulation is based on its incorporation into the structure of various transcription factors, including the zinc-finger transcription factors and hormone receptor proteins (Coleman, 1992; Blanchard et al., 2001; Dreosti, 2001; Cousins et al., 2003). Likely reflecting its role in gene regulation, zinc is required for the synthesis of a variety of enzymes and other proteins. Two key structural proteins, collagen and keratin, both require zinc for their synthesis (Underwood and Suttle, 1999).

Keratin is the major structural protein of the hoof horn, feathers, skin, beaks and claws, while collagen is the major structural protein of the extracellular matrix and connective tissues in internal tissues, including cartilage and bone. Decreases in collagen and keratin synthesis rates in zinc-deficiency can lead to a variety of defects including bone abnormalities, poor feathering, decreased tissue strength and dermatitis (Underwood and Suttle, 1999; Leeson and Summers, 2001).

Furthermore, collagen turnover rates are also decreased in zinc-deficiency, presumably because the collagenases/matrix metalloproteinases (MMPs) are zinc-dependent enzymes (Starcher et al., 1980; Pardo and Selman, 2005). Decreases in collagen synthesis and turnover rates would be predicted to cause decreases in tissue strength. This is important both for animal health and for post mortem processing. Decreased intestinal strength, for example, would be predicted to result in higher rates of intestinal breaks and condemnations during processing. Finally, zinc plays important roles in the development and proper functioning of the immune system (Shankar and Prasad, 1998; Fraker et al., 2000; Ibs and Rink, 2003). Deficiencies in zinc can lead to decreased immune function, as demonstrated by reduced T cell function, lower antibody titers and other deficits.

Like zinc, copper is essential for a wide variety of health and performance-related functions in all animal species. Functions performed by zinc are often enhanced by copper-dependent enzymes. For example, lysyl oxidase, the enzyme that crosslinks collagen subunits into mature protein forms to increase their strength, is copper-dependent (Rucker et al., 1998).

Lysyl oxidase also crosslinks the structural protein elastin, which is found in connective tissue, primarily in the cardiovascular system, intestines, and other tissues that change size as a consequence of fill. Because of its role in collagen crosslinking, copper promotes skin, bone, tendon and intestinal strength. Experiments in poultry have demonstrated that bone breaking strength correlates strongly with the extent of collagen crosslinking (Rath et al., 1999). In copper-deficient animals, therefore, the elastin and collagen may be unable to withstand the mechanical stresses typical of the cardiovascular or skeletal systems, respectively (O'Dell et al., 1961; Guenthner et al., 1978). Indeed, severe copper deficiencies have been reported to cause aortic rupture in multiple species, and bones may be fragile and easily broken (Guenthner et al., 1978; Opsahl et al., 1982; Underwood and Suttle, 1999). Like collagen, keratin synthesis requires zinc, and keratin crosslinking is copper-dependent.

Manganese is essential for growth and fertility of animals (Gallup and Norris, 1939; Underwood and Suttle, 1999). Furthermore, it plays a very important role in bone development, both in the embryo and after birth or hatch. The ground substance of developing bone, particularly the proteoglycan matrix in which collagen and elastin are embedded, requires Mn for glycosylation of its protein core molecule (Fawcett, 1994).

Proper development of this matrix is required for later stages of bone development. In a Mn-deficient animal, therefore, there can be a failure of endochondral ossification, resulting in chondrodystrophy and perosis (Underwood and Suttle, 1999). Evidence that Mn-deficiency causes these defects is shown by the fact that these two conditions can be corrected by manganese supplementation.

All three of these trace minerals play key roles in managing oxidative stress. Reactive oxygen species (ROS) generation is a normal byproduct of cellular energy production, and a primary weapon of the innate immune response (Mayne, 2003; Iqbal et al., 2004). Unfortunately, these ROS are damaging to cellular lipids, proteins and DNA and if left unchecked can induce a variety of undesirable consequences. The superoxide dismutase (SOD) enzymes form a first-line defence that converts oxygen radicals to hydrogen peroxide, which is a less toxic molecule (Hydrogen peroxide is then converted to water through the action of glutathione peroxidase, a selenium-containing enzyme). There are two forms of SOD in animal cells, the copper and zinc dependent form, in the cytoplasm, and the manganese-dependent form in the mitochondria (Underwood and Suttle, 1999). Decreases in SOD activity, for example in a mineral deficiency, can lead to increased amounts of lipid, protein and nucleic acid damage, which can induce cellular death (Rothstein et al., 1994; Troy and Shelanski, 1994; Kokoszka et al., 2001). The actions of SOD should not be underemphasized. Both the Cu/Zn- and Mn-dependent isoforms of SOD are two of only a handful of enzymes whose activity correlates with lifespan in simple organisms (Orr and Sohal, 1994; Parkes et al., 1998; Honda and Honda, 1999).

Zinc has been proposed to directly or indirectly manage oxidative stress in other ways as well, including through the induction of metallothionein, and as a required cofactor in the p53 transcription factor, which mediates the repair of DNA damaged by oxidative stress or other means (Ho and Ames, 2002; Formigari et al., 2007). Zinc deficiency has been shown in a variety of publications to increase oxidative stress in vitro and in vivo, as shown by increases in the prevalence of markers of oxidised lipids and damaged DNA, and the production of free radicals (Ho et al., 2003; Song et al., 2009). In zinc-deficiency, p53 loses its ability to bind DNA and promote repair, which can result in increased rates of apoptosis (Ho and Ames, 2002; Fraker, 2005). Indeed, zinc deficiency has been shown to increase cellular turnover rates in the small intestine (Cui et al., 1999; Richards et al., 2005).

Inorganic and Organic forms of Trace Minerals

Historically, zinc, copper and manganese have been supplemented in animal diets using inorganic salts such as oxides and sulfates. However, trace mineral salts tend to dissociate in the low pH environment of the upper gastrointestinal tract, leaving the minerals susceptible to various nutrient and ingredient antagonisms that impair absorption (and thus reduce bioavailability) (Underwood and Suttle, 1999). Antagonisms can occur between minerals. For example, high levels of zinc reduce the availability of copper, and the opposite is also true (O'Dell, 1989; Leeson and Summers, 2001; Zhao et al., 2008). In addition, phytic acid is able to form complexes with trace minerals that are very stable and highly insoluble, rendering the minerals unavailable for absorption (Oberleas et al., 1966; Leeson and Summers, 2001).

The phytic acid-mediated antagonism is amplified in the presence of calcium (Oberleas et al., 1966). A potential advantage of chelated trace minerals is that the binding of the organic ligands to the mineral should provide stability of the complex in the upper gastrointestinal system, thereby minimising mineral losses to antagonists and allowing the complex to be delivered to the absorptive epithelium of the small intestine for mineral uptake (Leeson and Summers, 2001). It should be noted that different organic trace minerals are not equally stable at low pH, and therefore will not necessarily increase the bioavailability of a given mineral to the same extent (Brown and Zeringue, 1994; Cao et al., 2000; Guo et al., 2001).

Measuring Trace Mineral Bioavailability

Measuring the deposition or storage of minerals into selected tissues (e.g. tibia or plasma zinc, liver copper, tibia manganese, etc.) is the most common output in trace mineral relative bioavailability (RBV) experiments (Underwood and Suttle, 1999). More recently, the use of mineral-responsive biomarkers, such as changes in gene or protein expression, or the activity of a mineral-dependent enzyme, has become more common (Payne and Southern, 2005; Richards et al., 2007; Huang et al., 2009; Richards et al., 2010).

Regardless of the response variables measured, it is surprising how many papers report little or no difference in OTM vs. ITM bioavailability. An extensive review of the zinc bioavailability literature in poultry, swine and other species was published in 1995 (Baker and Ammerman, 1995). In this review, the average bioavailabilities of a variety of inorganic (zinc oxide, zinc sulfate, zinc chloride, zinc carbonate, elemental zinc) and organic zinc sources (zinc lysine, zinc methionine, and zinc proteinate) were compared.

In poultry, relative to the sulfate, acetate or chloride forms of zinc (defined as 100% bioavailable in individual experiments), the average RBV of zinc methionine and a zinc proteinate was 125% and 100%, respectively (Baker and Ammerman, 1995; and references therein).

In swine, there were no differences between organic zinc (zinc methionine or zinc lysine) and inorganic zinc (zinc sulfate or zinc chloride) (Baker and Ammerman, 1995; and references therein). (Interestingly, zinc oxide ranged from 55% to 100% as available as zinc sulfate.) While this review is 15 years old, subsequent papers in the literature often report similar results. It seems likely that some OTMs truly will not be more bioavailable than ITMs, due to their inability to stay chelated or complexed in the low pH environment of the upper GI tract (Brown and Zeringue, 1994; Cao et al., 2000; Guo et al., 2001). On other occasions, however, true differences in bioavailability could be masked by experimental design.

Using tibia zinc content as the measure of bioavailability, Wedekind and colleagues have indicated that the bioavailability of zinc methionine relative to zinc sulfate ranges from 117% to 206% in broiler chicks, depending on the diet matrix (Wedekind et al., 1992). The differences in RBV were reduced in crystalline or semi-purified diets that contain low levels of antagonists such as fibre or phytic acid, and increased in corn-soy diets. Furthermore, these authors emphasized the importance of measuring RBV on the linear portion of the dose-response curve, rather than on the plateau.

Regardless of the response variables utilised, measuring the response on the plateau of the curve will minimise true differences in RBV. Indeed, in this same paper, Wedekind et al. reported no difference in zinc methionine vs. zinc sulfate RBV when supplementing at zinc levels above the tibia zinc breakpoint (Wedekind et al., 1992). Recent experiments with other chelated minerals support this finding. A study of zinc chelated by the methionine hydroxy analogue (HMTBa-chelated zinc, or MINTREX[R] Zn) performed on the linear portions of the dose response curves indicated that the zinc from this source was approximately 160% or 250% as available as the zinc from zinc sulfate, depending on the response variable measured (tibia zinc; or the small intestinal expression of the zinc responsive biomarker, metallothionein; respectively) (Richards et al., 2010).

Impact of Trace Mineral Supplementation on Tissue Development and Strength

As described above, zinc and copper play key roles in the synthesis and proper assembly of collagen. An experiment was conducted to test the effects of copper source and level on intestinal breaking strength (IBS) in broilers. Broilers were fed diets that were adequate for zinc, but were unsupplemented for copper (9 ppm Cu from ingredients), or supplemented with 25 ppm copper from copper sulfate, a copper proteinate, copper lysine, or HMTBa-chelated Cu (MINTREX[R] Cu). There was a significant improvement in IBS in all supplemented diets, but birds that were supplemented with HMTBa-chelated Cu had greater IBS than all other treatments. One likely explanation to understand these results is that the extent of collagen and elastin crosslinking in the control birds may have been low due to suboptimal copper status in the control diets. Addition of the various copper sources, but especially HMTBa-chelated copper, may have promoted collagen and elastin crosslinking beyond what occurred in the low-copper treatment. As described above, zinc and copper play important roles in bone development via their actions on collagen, while manganese-dependent enzymes promote formation of the proteoglycan matrix in the cartilage model for developing bone. Tibial dyschondroplasia (TD) is a common developmental defect in fast growing birds. In this condition, the cartilage model at the epiphysial plate fails to ossify, resulting in plugs of cartilage in place of true bone. Bones are therefore weak, and can result in bone breaks when the bird grows heavy. Rath and colleagues have demonstrated that tibias exhibiting TD have normal collagen content, but reduced amounts of sulfated glycosaminoglycans and MMP activity when compared to normal tibias (Rath et al., 1997).

These results suggest supplementation with manganese and zinc may alleviate TD. In a separate paper, this group also showed that the extent of collagen crosslinking in tibia correlated with bone breaking strength (Rath et al., 1999). These results imply an important role for copper. Given the high incidence of TD and bone weakness in fast growing meat birds, one would expect the incidence of these problems to be reduced with improved trace mineral nutrition. Indeed, recent commercial and university experiments in turkey flocks have demonstrated reduced TD, reduced lameness and increased bone breaking strength when the diets were supplemented with HMTBa-chelated trace minerals (Dibner et al., 2007; Ferket et al., 2009). It is interesting to note that in the control treatments, ITM levels were formulated at commercial levels, far exceeding published requirements. Thus, one might have predicted that these animals would not be deficient in trace minerals. Yet the physiological responses of these animals to chelated mineral supplementation indicate that the controls were deficient, at least with respect to supplying the minerals needed for optimal structural development.

Feeding chelated minerals can improve structural integrity of tissues even when fed at reduced inclusion rates. In a recent trial with 120,000 broilers, birds fed reduced levels of supplemental zinc, copper and manganese (32 ppm, 8 ppm and 32 ppm, respectively) as HMTBa chelates had significantly improved footpads relative to broilers fed much higher levels of zinc, copper, and manganese (100 ppm, 125 ppm and 90 ppm, respectively) as sulfates (Manangi et al., 2010). It is interesting to note that growth performance was not different between these two treatments, while trace mineral excretion was reduced in the birds fed chelates.

These data demonstrate that the HMTBa-chelated trace minerals met the requirement of these birds, even though they were supplemented at low levels.

Impact of Supplemental Trace Minerals on the Immune Response

As described above, trace minerals, especially zinc, are required for proper immune development and function (Shankar and Prasad, 1998; Fraker et al., 2000; Ibs and Rink, 2003). Deficiencies in zinc can cause decreased antibody responses to vaccination. Previous results with multiple inorganic (sulfate and oxide) and organic zinc sources (zinc methionine and HMTBa-chelated zinc) in poultry have demonstrated source differences in both immune development and response to antigenic challenge (Dibner, 2005; Moghaddam and Jahanian, 2009).

In each of these papers, supplementation with the organic zinc source enhanced cellular or antibody responses to vaccination. We wished to test the effect of feeding chelated trace minerals on the immune response to vaccination in pigs. Replacement gilts (50 per treatment) were fed diets supplemented with 165 ppm zinc, 16.5 ppm Cu and 38.6 ppm manganese, either as ITMs or an equal mixture of ITMs and HMTBa-chelated minerals. The pigs were vaccinated with a commercial vaccine for Mycoplasma hyopneumoniae on weeks 0 and 2 postweaning, and bled for antibody titers on weeks 0, 2, 4, 8 and 12. Titers were measured by a commercially-available ELISA. Log titers below 2.8 are considered to be negative titers according to the kit instructions. While both groups of pigs achieved a similar titer by 12 weeks, the gilts supplemented with the HMTBa chelates reached a positive titer 8 weeks prior to the gilts fed the control diet. These data suggest that for those eight weeks, the replacement gilts fed ITMs were not as protected against M. hyopneumoniae as the gilts fed the HMTBa-chelated minerals.

Oxidative stress results when the production of reactive oxygen species (ROS) exceeds the body's ability to detoxify the reactive species, or to repair the damage caused by them. Inappropriately high or chronic levels of oxidative stress can damage lipids, proteins and DNA, leading to high rates of cell death and turnover (Dibner et al., 1996; Girotti, 1998; Mayne, 2003). Poor oxidative stress management in production animals can result in decreased performance, compromised immune function, increased morbidity and poor meat quality (Sheehy et al., 1994; Buckley et al., 1995; Iqbal et al., 2004; Spears and Weiss, 2008). Although the roles that natural and synthetic antioxidants play in managing oxidative stress are well recognised, it is important to understand that trace minerals also participate in these processes (Underwood and Suttle, 1999; Ho and Ames, 2002; Formigari et al., 2007; Song et al., 2009). One method to estimate oxidative stress in a group of animals is to measure specific oxidised forms of lipids, proteins and nucleic acids from blood or tissue samples (Mayne, 2003). A battery study was conducted in broilers to investigate whether different inorganic and organic trace mineral forms could reduce oxidative stress. All treatments except the negative control were supplemented with 30 ppm zinc, 20 ppm manganese and 5 ppm copper. As an indicator of oxidative stress, the concentration of lipid hydroperoxides (LPO) was measured in plasma. Birds fed either the glycine-chelated trace minerals (GLYTREX[TM] chelated minerals) or the HMTBa-chelated trace minerals (MINTREX[R]) exhibited significantly ($p<0.01$) lower levels of lipid hydroperoxides in their blood versus the control, indicating lower oxidative stress in these birds. Inclusion of ITM or amino acid complexes did not reduce plasma LPO relative to the control ($p>0.45$) or each other ($p>0.2$).

Essential trace minerals such as zinc, copper and manganese play a wide variety of biological and physiological roles in animal development and health. These minerals take part in the antioxidant defence and DNA repair, bone and tissue development, and immune function. The importance of these minerals in animal agriculture is widely recognised, and virtually all diets are supplemented with these minerals. However, growing evidence supports the conclusion that the trace mineral requirements of production animals are not easily or consistently met by feeding inorganic forms of these minerals.

Well-designed experiments that investigate the relative bioavailability of trace mineral sources demonstrate that certain organic trace minerals can more effectively satisfy the trace mineral requirements of production animals. By providing a more bioavailable source of minerals, OTM supplementation has been shown to exert a variety of positive effects, including improved immune responses, enhanced bone and tissue development and strength, and reduced oxidative stress. Finally, due to their higher bioavailability, the requirement can be met at lower levels of inclusion in the diet, resulting in reduced excretion into the environment.

Type Broiler Breeders

It has been noticed in the USA that females reared with males often produce more eggs than females reared sex-separate. In order to understand this observation, a study (Mixgrow) was conducted to determine the effect of mixing males with females at different ages. Yield males were fullfed on an 18% CP diet until mixed with females at two, four, six, or eight weeks of age. The yield females received an 18% CP diet for one week followed by a 15% CP diet to photostimulation. The female feeding programme used was one that had been shown to be successful for the "standard" type of broiler breeder pullet. Female body weights were virtually identical across male treatments. These fertility numbers are lower than optimum because males and females were fed together after 21 weeks of age to exaggerate the effect of cumulative nutrition during rearing and to allow the males to be exposed to a decreasing feed allocation after 35 weeks of age. In spite of this, some of the pens with the eight-week mixed males exhibited fertility in excess of 90% at 64 weeks of age without any body weight control or separate feeding.

The later mixing age males (six and eight weeks) were more resistant to the feed reduction after peak egg production because they reached sexual maturity with a greater nutrient reserve. The actual feed intake of the males mixed with females at six weeks of age (as an example) and that of the females can be estimated from the body weights taken from all birds every two weeks using the formulas of Combs (1968). The males consumed about 125% to 150% of the female feed intake depending upon age when mixed and body weight. This would give an actual cumulative ME intake of over 34,000 kcal and 1600 grams of CP for both the six week and eight week mixed males. This agrees with other data from our laboratory with separate-grown males. The data also show that the real pattern of female feed consumption differed significantly from the programmed pattern, especially after 14 weeks of age.

This must be extremely important as females that were grown sex-separate on the programmed female feed amounts laid ~35 fewer eggs per hen. These data (and field experience) suggest that larger feed increases late in rearing (in blackout where there is little reproductive development) for "yield-type" pullets results in excessive body weight and excessive "fleshing" (breast meat development). Much has been said about the need for good "fleshing" in "standard" strains of parent stock but the situation is much different for the "yield-type" pullet. Excess breast meat appears to reduce egg production.

We must be careful to not give too much feed too early (before onset of lay) as we may simply increase female body weight, primarily breast meat, and cause reproductive problems such as peritonitis. The excess breast meat probably increases maintenance and inhibits reproductive development.

This may be why heavy breasts relative to fat pad develop when feed increases are too rapid in "yieldtype" females. These birds with excess breast meat relative to fat pad tend to exhibit a reduced appetite in hot weather (even in tunnel-ventilated and evaporatively cooled houses), increased susceptibility to heat stress, poor peak egg production and lay poorly thereafter. A conservative feeding approach both before and after photostimulation would be advisable with "yield" females until one becomes familiar with the particular strain of broiler breeder in the specific situation. It is better for the hens to be late coming into production than to exhibit high mortality and poor egg production. These problems are uncommon with a "standard" type broiler breeder hen.

In a manner similar to the need to modulate any large increases in feed intake, diets should be formulated to minimise abrupt changes in composition that will create situations that are similar to abrupt changes in the feeding rate. A single dietary ME for all diets is recommended to assist production managers maintain consistent feed increases.

Similarly, modern broiler breeders may respond robustly to abrupt changes in protein with an unexpected increase in body weight. A smooth transition among starter-grower-breeder diets or starter-grower-prebreeder-breeder diets should be considered during feed formulation. It is suggested that total lysine levels be ~5% of crude protein and methionine + cystine be ~0.60-0.63% of the diet for most feeds. It is probable that "yield-type" females perform better with a slightly lower protein breeder feed than can be fed successfully to a "standard" female. A 16% CP diet with ~0.80-0.82% total lysine should be sufficient to support egg production without producing excessive amounts of breast meat.

Broiler Breeder Males: Dietary Protein and Metabolisable Energy

Few data exist that link intake of ME during rearing to breeding performance. However, the findings of Vaughters et al. (1987) indicated that a relationship between ME consumed during rearing and fertility may exist. Our data suggest a minimum cumulative ME intake of ~30,000 kcal prior to photostimulation. However, most data suggest that reproductive fecundity is directly related to daily ME intake during the breeding period and that daily ME intake should somehow be proportional to body weight and body weight gain. It should be stated that Parker and Arscott (1964) and Sexton et al., (1989b) observed that decreased fertility was preceded by decreased dietary ME intake during the breeding period. In cages, Attia et al. (1995) fed Ross males 300, 340, or 380 kcal ME per day. They found no fertility differences, but did note increasing testis weights with increasing ME intake. In floor pens from 26 to 60 weeks of age, Attia et al., (1993) found the 300 kcal ME males to weigh less and have lower fertility than the males consuming 340 and 380 kcal ME per day. These data clearly show a differential effect of ME intake in cages versus floor pens due to the difference in relative activity levels. All the birds in cages probably received enough ME to satisfy their reproduction requirements. However, in the floor pens, it appeared that the birds on the lowest ME intake did not receive enough nutrients for reproduction due to the increased maintenance requirement required for increased activity.

It is also very interesting that these authors found a dose-related decrease in 42-day broiler weights with decreasing ME allocation to the breeder males. Presumably, these data suggest that males that have the potential to produce the largest broilers require more ME to breed in natural mating conditions. These data also suggests that excessive efforts to control male body weight can reduce broiler performance.

Confusion about optimum diets for males began when Wilson et al. (1987a) fed 12%, 14%, 16%, and 18% CP diets to males from four to 53 weeks of age. The 10 males used per treatment were placed in cages at 14 weeks of age. There was no lighting programme detailed in the manuscript and is presumed to be natural daylight during rearing with artificial supplementation at some unspecified point. Cumulative CP to 21 weeks was 1220 grams and 1385 grams, respectively for the 12% and 14% groups. This total increased to 1650 grams at 27 weeks of age for the 12% group, the time of the first artificial ejaculations in this particular study.

By comparison, males in natural mating conditions need to mature by ~22 weeks of age for best results. No significant differences in semen volume, testis weights, and spermatozoal concentration among the diets were found, but significantly more males produced semen as a result of abdominal massage on the 12% and 14% CP diets. Although there were no significant differences in body weight among the treatments, the 12% and 14% males did exhibit a generally more consistent body weight gain throughout the breeding period. It is important to note that all the diets used in this and subsequent studies from this laboratory at Auburn University had total lysine as 5.1% to 5.3% of total CP and total methionine + cystine as 75% to 77% of lysine in corn-soy based diets. This was similar to the dietary approach used by our laboratory at North Carolina State University, but may differ somewhat from observed commercial practice where low protein male diets are often not properly balanced. We like to have lysine as 5% of CP and methionine + cystine in the range of 75% to 83% of lysine.

In a recent study from the same laboratory at Auburn University, Zhang et al. (1999) made a comparison of 12% and 16% CP diets from four to 52 weeks of age. As in previous reports, there was a higher percentage males producing semen as a result of artificial ejaculation, but there were again no differences in semen quality or quantity. Given that differences in semen quality or quantity are not usually found as a result of difference in CP intake, one has to question if the reported higher percentage males producing semen as a result of artificial ejaculation is simply an artifact of the semen collection process with birds that may vary in body conformation.

This response (percentage males producing semen) seems to consistently take the form of a dose response while all other variables show no such dose response. In the experiment of Zhang et al. (1999), the daily ME allowance was 325 kcal during the breeding period. As shown later, this energy allocation is slightly low. A gradual decline in semen production with increasing age and body weight was observed, irrespective of CP level of the diet. The authors interpreted this to mean that continued body weight gain was necessary to maintain optimal male reproductive function. Continued body weight gain clearly would require appropriate increases in ME allowances as body weight increased.

The extensive French work, led by de Reviers showed that heavy weight line males exhibit greater problems with persistency of testes size and semen production when compared to medium weight male lines. Photostimulation of heavy weight line males typically result in a robust, but short, response in testicular weight and semen production while medium weight male lines exhibit better persistency of these traits. It is presumed, as no nutritional data were given in these reports, that both male lines were fed typical low-density diets. The problem of lack of persistency of semen production can be solved, if one is using artificial insemination, by simply not photostimulating the birds and allowing the males to reach sexual maturity at their own pace, presumably after consuming sufficient nutrients.

Therefore, if a bird were deficient in CP during the growing period the effects would be most noticeable around the onset of sexual maturity. Vaughters et al., (1987) fed diets containing 12%, 15%, or 18% from 24 to 27 weeks of age (early breeding period) and reported initial fertility to be highest for the 18% CP diet in natural mating conditions. This suggests a relationship between sexual development and the initiation of reproductive function. Turkey and broiler breeder hens are both known to exhibit an intense desire to mate prior to the onset of egg production.

When turkey hens were inseminated during this period of prelay receptivity, there was a significant increase in life-of-flock fertility even in the presence of marginal spermatozoal numbers (McIntyre et al., 1982). This early mating presumably leads to enhanced spermatozoal storage. This may also be true for broiler breeders. It is clear that broiler breeders that exhibit low initial fertility under commercial natural mating conditions, where sexual maturity is needed at about 22 to 24 weeks of age, have difficulty achieving optimum fertility at later ages.

Although there appears to be an impact of CP during the growing period on fertility during the breeding period, dietary CP appears to have less impact during the breeding period. Diets from 5% to 16.9% CP have produced similar results in cages. The reason that these previous workers did not see more differences in fertility due to breeder dietary CP was probably due to the fact that their experiments were often initiated later in the breeder period (after 28 weeks of age).

In these experiments, it appears that the birds were not marginal in CP before the experimental diets were applied, which made it difficult to detect fertility differences due to differences in breeder dietary CP. These data also suggest that low protein male feeds should not be used before sexual maturity is complete.

Data from our laboratory suggest the minimum cumulative CP intake required prior to photostimulation for broiler breeder males involved in natural mating to be on the order of 1600 grams, as compared to the 1200 grams required for female. We have found that it is possible to achieve this nutrient target with diets ranging from 12% CP to 17% CP. Moreover, our data, shown below, demonstrate the interaction of body weight and feeding programme that influence male reproduction so profoundly. The broiler breeders were the Ross 308 package but the data are illustrative of our data with Cobb 500 and Arbor Acres Yield broiler breeder packages as well. All of these birds were reared separately from the females and fed sexseparate during the breeding period. The effect was more pronounced for the 17% CP males that were slightly larger and evidently less resistant to the imposed feeding deficiency. The problem was corrected by a five grams increase in daily feed allocation for the males. The cumulative intake of nutrients at 21 weeks of age were 1568 g CP and 36,593 kcal ME for the 12% males and 2123 g CP and 36,593 house was at 23 weeks of age. Again, the data suggest that if the minimum nutrition is adequate, it is not important what dietary protein level is used to achieve the goal. The males with the most consistent body weight gains produced the best fertility.

Body Weight

It has long been clear that feed restriction to control body weight is both obligatory and beneficial in broiler breeders. However, excessive feed restriction of males during part or all of the growing period has been associated with decreased early fertility (Lilburn et al., 1990). Based upon the discussion above, it is thought that this effect is due to insufficient cumulative nutrition at photostimulation.

The major impetus for sex-separate feeding during the breeding period was the observation that poor fertility was associated with overweight males and separate feeding was believed to be necessary to control male body weight. However, caged males fed near ad libitum are known to exhibit excellent spermatozoal production. This suggests that an appropriately controlled feed allocation rather than severe restriction is required. It is likely that overly severe feed restriction has actually caused fertility problems due to reduced mating activity as a result of caloric deficiencies. This may help explain the observations of Hocking (1990) who performed experiments with males in floor pens with natural mating during the breeding period. He found a curvilinear relationship between body weight and fertility. This implied that if body weight were too low or too high there would not be optimum fertility. He observed that underweight males were not physiologically sufficient while overweight males often were physically incapable of completing the mating process. He suggested an optimum body weight for optimum fertility that changed with age. He concluded that restricted control of body weight should allow an increase in body weight with age of the male.

We conducted a study to examine this inconsistency. We found that a decrease in fertility coincided with a decrease in female feed allocation and an increase in male body weight in situations where males were fed with females. In a similar manner, a decrease in male feed allocation in situations where males and females were fed separately caused a transient decrease followed by an increase in male body weight coincident with a decrease in fertility. Thereafter, fertility again increased when the feed allocation was increased in the separate-fed males. Male body weight was better controlled and fertility improved when the male feed allocation was increased slowly rather than decreased.

What can be the explanation for the paradox shown above? As an example, a typical male at ~30 weeks of age will weigh ~4.00 kg (8.8 lbs.). The daily maintenance requirement at ~21°C (70°F) is ~306 kcal while that of a 4.45 kg (9.8 lbs.) male at ~45 weeks of age would be ~329 kcal. Unless there has been an increase in daily feed allocation proportionate with the body weight gain, the 4.45 kg male would have to exhibit negative growth (lose body weight) as the male mobilised body reserves to make up the energy deficiency.

This would continue until the energy reserves of the larger male were exhausted. At this time, mating activity would decrease as testosterone levels decreased. The male would then gain body weight because of inactivity. This could lead one to conclude that males do not necessarily cease mating because they gain excessive body weight, but that males gain excessive body weight because they cease mating! In the same way that our best egg production occurs when the females slowly gain body weight, our best fertility occurs when the males slowly gain body weight. As the male does not exhibit a decline in daily energy requirement as does the female (due to decreasing egg production) it is suggested that the daily feed allocation be increased at least one gram every three to four weeks during the breeding period such that the male body weight increases slowly but consistently and remains within limits established by practical experience and known to be associated with good fertility. Ken Krueger (1977) found that male turkey semen production could be maximised for the entire life cycle by maintaining the toms on a feeding regimen that supported a consistent weekly body weight gain. Any loss in body weight was associated with a decline in semen production.

Broiler Breeder Male Mortality

The mortality of "yield-type" broiler breeder males during the laying period has become a costly problem for the USA poultry industry. The average male mortality from 22 to 64 weeks during the years 1995 to 1999 was approximately 43% (AgriStats, Inc., 6510 Mutual Drive, Fort Wayne, IN 46825). The cause of the majority of this mortality is unknown. To test a theory about the cause of this high mortality, Ross 308 females and non-dubbed Ross males were raised sexseparate on either a "linear" or a "concave" feed allocation programme. One group of males was reared with females on a "mixed" programme. Birds were grown on a daily 8-hour light and 16 hour dark lighting programme and both feed and water were controlled. At the end of 21 weeks, the birds were moved to a curtain-sided laying house and photostimulated. "Linear" grown males received constant feed increases of 2.4 g per male/week from four to 28 weeks. After 28 weeks, males received a constant feed amount of 117 g (342 kcal ME) per bird. All separate grown males received the same amount of cumulative feed through 21 weeks that resulted in a cumulative CP intake of 1600 g and a cumulative ME intake of 32,000 kcal per male at photostimulation at 21 weeks of age.

Separately grown males on the "linear" programme had significantly higher body weights than separately grown males on the "concave" programme. However, there were no differences in body weight due to treatment after photostimulation (22 weeks). All males had similar body weights at 22, 26, 28, 40, and 52 weeks of age.

It appeared that the "mixed" grown males had less mortality during this period although these differences were not statistically significant. Males grown separately on the "linear" programme or "mixed" with females had significantly higher mortality from 30 to 44 weeks when compared to males grown separately on the "concave" programme. From 45 to 64 weeks, "linear" males numerically had the highest mortality with "mixed" and "concave" males having similar mortality. When mortality was compared from 30 to 64 weeks, "concave" males had significantly lower mortality when compared with the "linear" males. "Mixed" males were intermediate. This same trend was observed overall (22-64 weeks). All data indicated that males grown on a "linear" feed allocation programme exhibited higher mortality than males grown on a "concave" feed allocation programme. It appears that the majority of the mortality due to "linear" feeding can be expected to occur between 30 to 44 weeks of age.

It appeared that 117 g of feed (342 kcal ME) per male per day was adequate to keep males grown on a 17% CP diet slowly gaining weight from 28 to 60 weeks of age under our current research management that utilises strict male and female exclusion grills. Fertility was excellent with these males. However, under commercial conditions, a gradual increase in male feed allocation would be advised. Therefore, it appears that the feed allocation programme used during the growing period in association with the time of photostimulation can influence broiler breeder male mortality. It appears that this occurs irrespective of body weight. The various groups of males employed in these experiments exhibited average body weights that were not remarkably different. Thus, one can conclude that management of feeding programmes should take limited precedence over body weight management.

Overview of Separate Male Rearing

Males may successfully be reared separately from females throughout the growing period. Careful attention to the feeding programme must be exercised as demonstrated by the field observations outlined below. There has been much discussion about optimum male BW at four and 20 weeks of age. A thorough examination of all available data suggest a minimum required cumulative nutrient intake from day old to photostimulation of ~1600 g CP and ~32,000 kcal ME per male and a specific feeding programme approach is required to minimise mortality and maximise fertility.

We can conclude that the amount of nutrients a bird has available throughout its life impacts fertility and egg production. Metabolisable energy (or feed allocation) available to the bird during the breeding period is directly correlated to fertility, egg production and body weight. Protein accumulated in the bird during rearing influences the age of sexual maturity and the level of initial fertility for both males and females. Dietary CP has the largest impact during the grower and pre-breeder periods, as this is when most of the CP required for initial sexual development is accumulated.

Body weight and house temperature need to be controlled within certain limits throughout the life of the flock, however temperature and body weight management is most critical late in the breeding period because body weight is greatest at this time. The data clearly show that no specific diet has more or less utility for a male broiler breeder. Diets ranging from 12% CP to 17% CP can be fed provided that the cumulative intake of CP to photostimulation is sufficient to support initial sexual development.

However, it is clear from practical experience that changing from a moderate or high CP feed to a low CP feed prior to sexual maturity has adverse effects on broiler breeder fertility. It is most important to maintain consistent body weight gain throughout the life of the broiler breeder. Abrupt increases or decreases in body weight are clearly associated with changes in fertility and egg production. This infers a need to closely align ME intake to maintenance requirements that are driven by body weight and temperature. In summary, all the rules for the "standard" broiler breeder remain basically true, but more attention must be paid to these details. The most obvious exception to the basic rules is that excessive "fleshing" can be detrimental in the yield-type female because it can increase sensitivity to environmental temperature, reduce egg production and appetite, and increase mortality.

Special Notes and Acknowledgements

Some estimates of metabolic energy requirements in the text were based upon the formulas of G. F. Combs, 1968, the Proceedings of the Maryland Nutrition Conference for Feed Manufacturers. These estimates have been found to be reasonably accurate, but may need to be adjusted slightly for strain and age effects and should be used with some caution.

Portions of this manuscript were excerpted from the Proceedings of the Poultry Beyond 2005 Conference held in Rotorua, New Zealand in February 2001 and from the Proceedings of the Australian Poultry Science Symposium held in Sydney in February 2001.

Poultry Broiler Farming

Poultry meat is an important source of high quality proteins, minerals and vitamins to balance the human diet. Specially developed breeds of chicken meat (broiler) are now available with the ability of quick growth and high feed conversion efficiency. Depending on the farm size, broiler farming can be a main source of family income or can provide subsidiary income and gainful employment to farmers throughout the year.

Poultry manure has high fertilizer value and can be used for increasing yield of all crops.

i) The advantages of broiler farming are

ii) Initial investment is a little lower than layer farming

iii) Rearing period is 6-7 weeks only

iv) More number of flocks can be taken in the same shed

v) Broilers have high feed conversion efficiency i.e. least amount of feed is required for unit body weight gain in comparison to other livestock.

vi) Faster return from the investment

vii) Demand for poultry meat is more compared to sheep/Goat meat.

Scope for Broiler Farming and its National Importance

India has made considerable progress in broiler production in the last two decades. High quality chicks, equipments, vaccines and medicines are available. Technically and professionally competent guidance is available to the farmers. The management practices have improved and disease and mortality incidences are much reduced. Many institutions are providing training to entrepreneurs. The broiler population has increased from 4 million in 1971 to 700 million in 1998. An average annual growth rate of 20% was estimated during the eighth five year plan (1992-1997). Increasing assistance from the Central/State governments and poultry corporations is being given to create infrastructure facilities so that new entrepreneurs take up this business. Broiler farming has been given considerable importance in the national policy and has a good scope for further development in the years to come.

Financial Assistance Available from Banks/NABARD for Broiler Farming

NABARD is an apex institution for all matters relating to policy, planning and operations in the field of agricultural credit. It serves as an apex refinancing agency for the institutions providing investment and production credit. It promotes development through formulation and appraisal of projects through a well organised technical services department at the Head Office and technical cells at each of the Regional Offices.

Loan from banks with refinance facility from NABARD is available for starting broiler farming. For obtaining bank loan, the farmer should apply to the nearest branch of a commercial or cooperative or regional rural bank in their area in the prescribed application form which is available in the branches of financing bank. The technical officers attached to or the manager of the bank can help or give guidance to the farmers in preparing the project report to obtain bank loan.

For poultry farming schemes with very large outlays detailed project reports are required to be prepared. The items of finance would include construction of broiler sheds and purchase of equipments. Cost of one day old chicks, feed, medicine and labour cost for the first 7 weeks period for the first cycle, are also considered. Facilities such as land development cost, fencing, water and electricity, essential servants quarters, godowns, transport vehicles, broiler dressing, processing and cold storage facilities can also be considered for providing loan. Cost of land is not considered for loan. However, if land is purchased for starting a broiler farm, its cost can be treated as party's margin money upto 10% of the total cost of project.

Scheme Formulation for Bank Loan

A scheme can be prepared by the beneficiary after consulting local technical persons of State veterinary department, poultry corporation or private commercial broiler hatcheries. If possible, they should also visit the progressive broiler farmers in the area and discuss the profitability of farming. A good practical training and experience on a broiler farm will be highly desirable, before starting a broiler farm. As broilers have to be sold after attaining 6-7 weeks of age, a regular and constant demand for broiler meat and nearness of the farm to the market should be ensured. The scheme should include information on land, water and electricity facility, marketing aspects, training facilities and expertise of entrepreneurs and the type of assistance available from State government, poultry corporations, local hatcheries. It will also include data on proposed capacity of the farm, total cost of the project, margin money to be provided by beneficiary and requirement of bank loans, estimated annual expenditure, income and profit and the repayment of loan and interest.

Requirements of Good Project

The bank officers also can assist in preparation of the scheme or filling in the prescribed application form. The scheme so formulated should be submitted to the nearest branch of bank. The bank will then examine the scheme for technical feasibility and economic viability.

Technical feasibility-this would briefly include :

(a) Suitability of climate and potentiality of the area

(b) Technical norms including schedule for replacement of flocks

(c) Facilities and infrastructure available for supply of inputs, veterinary aid, marketing, training/experience of the beneficiary.

Financial viability-this would briefly include :

(a) Unit cost and loan requirement

(b) Input costs for chicks, feed, veterinary aid, labour and other overheads

(c) Output costs i.e. sale of broiler for meat, manure and other miscellaneous items.

(d) Calculation of annual gross surplus (income-expenditure)

(e) Cash flow analysis

(f) Repayment schedule i.e. repayment of principal loan amount and interest.

Other documents such as loan application forms, security aspects, margin money requirements etc. are also examined. A field visit to scheme area is undertaken for conducting techno-economic feasibility study for appraisal of the scheme. The model economics of Broiler farming unit of 4000 birds is given in Annexure IIa to IIf.

Sanction of Bank Loan and its Disbursement

After ensuring technical feasibility and financial viability, the scheme is sanctioned by the bank. The loan is disbursed in kind in 2 or 3 stages, such as against the creation of specific assets, construction of sheds, purchase of equipment and machinery, recurring cost on purchase of chicks, feeds, medicines, etc. The end use of the loan is verified and constant follow up is done by the bank.

Vegetable-based Feed Formulation on Poultry Meat Quality

Livestock production plays an important role in the agricultural sector of every nation particularly in the West African sub-region. The satisfactory outcomes of agricultural activities depend, to a large extent, on the use of feeds that are safe and of high quality. Livestock industries usually formulate feeds from materials that are either edible or inedible by man. These feeds, when ingested by the animals, enhance the animal productivity in terms of number and nutrient quality, to meet most of the immediate nutrient requirements of man.

World feed resources are on the verge of rapid decline, caused probably by increase in the number of humans and human activities. Hence, it is inevitable that conventional animal feeds should become increasingly more expensive. This has led to a search for new, often unconventional feeds, and effective methods of processing presently inedible roughages into more acceptable and nutritious feeds.

Agricultural practice in West Africa, and indeed most developing countries, consists of small- scale farming. The farmers in this sub-region have, in general, low level of agricultural education and at the same time are handicapped by insufficient capital. According to Payne and Wilson, the unavailability of capital and increasing worldwide cost of energy, purchased feeds, equipment and pharmaceuticals may in the long run delay or even halt the complete industrialisation and urbanisation of poultry production in tropical countries. Under such circumstances, subsistence and small-scale production methods with additional improvements may become relatively attractive to this sector of the population.

Feed ration (formulation) involves combining different ingredients in proportions necessary to provide the animal with proper amounts of nutrients needed at a particular growth stage. The ration should be palatable to the animals and not cause any serious digestive disturbances. Different species of animals have different requirements for energy (carbohydrate and fat), proteins, minerals and vitamins in order to maintain functions like homeostasis, reproduction, egg production, lactation and growth. Feed formulation does not merely involve mathematical calculations but factors such as cost, presence of anti-nutritional factors, texture, moisture, processing, digestibility and acceptability to the animal. One of the most important roles of animal production is to provide high quality protein for human consumption; to achieve this, animals should be fed correct proportions of high quality protein. Broilers are fast growing birds, which mature at 8 to 10 weeks; they are tender-meated with soft, pliable and smooth textured skin depending on feed type and management. A targeted live weight of 1.8 kg is attained at 12 weeks. There are two types of broiler rations, namely, the broiler starter mash fed from day one to fourth or fifth week and the broiler finisher mash, fed from week 4 or 5 until slaughter.

Today, poultry industries are highly commercial and nutrition is by far the most important single factor accounting for 65-75% of the inputs in the industry. Currently, most poultry are given fat-rich meals which impact negatively on the quality of products and in turn affect humans after consumption. However, poultry farmers have been trying several alternatives of feed formulation to enhance feed quality on meat and egg production. One of the major problems encountered in the tropics especially by small-scale farmers who wish (or are forced by lack of funds) to formulate their own rations is inadequate knowledge of poultry nutrition.

When poultry diets are designed using conventional foodstuffs, they follow a fairly predictable pattern and approximate quantities of the various ingredients. Another handicap for these local small poultry farmers is the inaccessibility of the vitamin and mineral premix and lack of data on the average nutrient content of the many local foodstuffs in the ration. Vegetable-based feeds are a rich source of essential plant amino acids, vitamins, minerals and antioxidants. Further to the rich contents mentioned, it has been established that green vegetable leaves are the cheapest and most abundant source of proteins because of their ability to synthesize amino acids from a wide range of available primary materials such as water, carbon dioxide and atmospheric nitrogen. This study, therefore, investigated the effect of vegetable substitute for vitamin pre-mix, lysine and methionine in commercial broiler starters' mash on the fat and protein content of meat production.

Materials and Methods

The experiment was carried out at the poultry unit of the Department of Agriculture, Babcock University, Ogun State, Nigeria. Fifty day-old chicks were purchased from Joy Veterinary Services and randomly distributed into two groups: control and experimental groups. The two groups were fed and given water ad libitum throughout the experiment. The control groups were fed commercial broiler starters' mash while the experimental groups were fed vegetable-based formulated feed. Slurry of 10kg of pawpaw and 10kg of banana was made by blending them into pulp of uniform mixture. The slurry was mixed with grounded maize, soya bean cake, groundnut cake, fish meal, palm kernel cake, bone meal, oyster shell and salt (NaCl). The mixture was sun- dried for 48 h. Ten kg of pumpkin leaves were also sun-dried for 48 h and milled together with the dried slurry mixture to form the experimental diet.

It is important to note that there are several methods of feed formulations including the square method, simultaneous equation method, two- by- two-matrix-method, trial- and- error method and linear programming method.

For this study, trial-and-error method of feed formulation was adopted. Birds that were cared for 6 weeks were administered glucose, vitamins and antibiotics on the first day of experiment. Gumboro vaccine was administered on the 7th day while from the 19th to the 20th day of the experiment, the birds were given coccidiostat. Water was given twice daily to the birds only for the last 6 days prior to slaughtering. Six (6) birds were randomly selected from each group and slaughtered. The protein concentration was determined according to Lowry et al. using bovine serum albumin (BSA) as standard, while the percentage crude protein, moisture and ash content of organs including the head, gizzard, heart, lung, small intestine, large intestine, upper limbs, lower limbs and liver were determined according to Pearson's Chemical Analysis of Foods. Live performance of broilers in terms of muscle protein weight and body weight gain was assessed according to Baker and Chung and Degussa.

Statistical Analysis

All data obtained from determinations of percentage crude protein, moisture ash and fat content in vegetable and standard formulated feeds as well as plasma and muscle protein, weight of organs and fat contents of experimental and control birds were subjected to SPSS for Windows version 15.0 statistical package. Comparison of means was done using paired sample t test. P value less than 0.05 was considered to be significant. Data were reported as mean [+ or -] standard deviation.

The vegetable-based feed formula contained a significantly (P<0.05) higher crude protein (15.75 [+ or -] 0.14%) and moisture content (23.3 [+ or -] 2.36%) than those of standard commercial feed formula crude protein (9.63 [+ or -] 0.13%) and moisture content (16.7 [+ or -] 2.23%). The ash content (10.0 [+ or -] 4.08%) and fat composition (2.5 [+ or -] 0.78%) were the same in both feed formula. There was no statistically significant difference (P>0.05) in organ weights (head, gizzard, liver, heart, lungs, small intestine, large intestine, upper limbs and lower limbs) between experimental and control groups. Fat analysis indicated a significant ($P<0.05$) decrease in the serum total cholesterol and mean fat composition in the serum, heart, gizzard and muscles of the experimental groups than the control group. There were, also, no significant differences ($P>0.05$) in the plasma protein and muscle protein contents between experimental and control groups. However, the vegetable-based fed birds had a higher muscle protein weight and body weight gain than commercial starter fed bird.

The study indicated that low fat and high protein meat can be obtained from birds fed with experimental vegetable feed than those fed with commercial broiler's starter mash. Therefore, the vegetable-based products serve as a source of the essential ingredients required by poultry farmers during the formulation of broilers' feed. The vegetable feed formula may enhance the poultry meat products in terms of nutritive value that would in the long run be beneficial to the health of meat consumer's and possibly serve as a source of economic value to the poultry farmers.

It is recommended that the addition of vegetable-based products in poultry feed formula would serve as a cheap source of amino acids, antioxidants, vitamins and bioactive metabolites, that are necessary for the growth and development of broiler birds. The production of such birds could in turn improve the health status of meateaters by boosting their immunity.

This would contribute greatly towards the prevention and reduction of neurodegenerative diseases associated with lipid-rich diet. Hence, the adoption of green vegetables and vegetable pulp substitutes for pre-mixed vitamins during feed formulation in West Africa and other developing nations, would serve as a cheap and natural source of ingredient in poultry feed formulation for small scale farmers.

Comparative Effects of Phytase Derived from Escherichia Coli and Aspergillus Niger in Sixty Eight-week-old Laying Hens Fed Corn-soy Diet

Phosphorus has been identified as an essential mineral in the formation of eggshells and the metabolism of laying hens (Sohail and Roland, 2002). Phytate is the primary storage form of phosphorus found in plant feed ingredients; however, phytate P is poorly utilised due to the limited inherent phytase activity in the gastrointestinal tract of non- ruminants (Adeola et al., 2004). Previous studies have shown that supplemental phytase exerts beneficial effects in poultry

(Rutherfurd et al., 2004; Francesch et al., 2005), cheaper and more effective sources of phytase would hasten the use of phytase in the poultry industry.

Phytase derived from different sources may differ in biochemical and biophysical properties such as the optimum pH and the ability to resist hydrolysis within the digestive tract, which will affect the ability of the phytase enzyme to function effectively and consistently (Onyango et al., 2005). It is well expected that phytase derived from Escherichia coli (ECP) will have superior results when compared to phytase derived from Aspergillus niger (ANP) due to differences in their biochemical and biophysical properties (Rodriguez et al., 1999b; Wodzinski and Ullah, 1996). However, the results of practical feeding studies have been inconclusive. For example, Augspurger and Baker (2004) and Rodriguez et al. (1999a) found that ECP had a higher efficacy when compared to ANP in broilers and pigs, whereas similar efficiancies of both phytases were observed when they were fed to young chicks (Leeson et al., 2000) at the same dosage. To the best of our knowledge, previous studies conducted to evaluate the effects of phytase supplementation on hens have primarily been conducted during various times in the first cycle of laying hens, but limited information exists regarding the effects of phytase supplementation at 68-weeks old fed corn-soy diets. Besides, experiment conducted by Boling et al. (2000) also shown that older hens are more sensitive to P deficiency, which may provide a proper environment to compare this two phtyase. One commercial phytase enzymes and an experimental E. coli-derived phtyase (ECP) were used in these experiments. Natuphos (NAT) is a recombinant enzyme from Aspergillus niger that is classified as a 3-phytase, with hydrolysis of the phosphate moiety being initiated at the 3-position on the phytate molecule. Phytase Optiphos (OPT) was isolated from Escherichia coli, which was expressed and produced through a yeast expression system that has been described previously (Rodriguez et al., 1999a, b). Prior to use in this experiment, the activities of the purified phytases were determined based on the release of inorganic P from sodium phytate in 0.2 M citrate buffer as previously described (Han et al., 1999). One phytase unit (FTU) was defined as the amount of enzyme required to liberate 1[micro]mol of inorganic P per min from 0.0015 mol of sodium phytate at 37[degrees]C with a pH of 5.5.

Feeding Regimen

A total of two hundred and sixteen 68-week-old Hy-Line brown laying hens (Yang Ji Hatchery, Cheonan, Choongnam, South Korea) were evaluated in 6-week feeding trial. The hens were randomly allotted into one of six treatments according to the individual initial BW, with six replications per treatment. Each replication consisted of three adjacent cages, with two hens per cage. In addition, the replications were equally distributed into the upper and lower cages to minimise the effect of cage level. The hens were housed in a windowless laying house under a 17 h light: 7 h dark photo period at approximately 21[degrees]C. There was a 7 d adjustment period prior to the start of the experiment, during which the hens were provided with PC diet. All cages were equipped with nipple drinkers and common trough feeders and feed and water were provided ad libitum throughout the experimental period. The animal care protocol used for this experiment was approved by the Animal Welfare Committee of Dankook University.

Experimental Design and Diets

Diets were formulated to have the same nutrient density, except the levels of P differed. Hens were provided with one of the following dietary treatments: i) Positive Control (PC; available phosphorus (AP) 0.4%); ii) Negative Control (NC; AP 0.2%); iii) NAT1 (NC+250 FTU/kg NAT); iv) NAT2 (NC+250 FTU/kg NAT); v) OPT1 (NC+250 FTU/kg OPT); vi) OPT2 (NC+250 FTU/kg OPT). The PC diet was formulated to meet or exceed the NRC (1994) nutrient requirements for laying hens. The NC diet was formulated the same as the PC, except that the AP level was reduced to maximise the response to enzyme supplementation.

Sample and Measurement

Parameters of production performance and egg quality : Daily records of egg production and weekly records of feed consumption were kept throughout the experimental period. The egg production was expressed as an average hen-day production, which was calculated from the total number of eggs divided by the number of days and summarised on an average basis. In addition, a total of 30 salable eggs (no shell defects, cracks, or double-yolks) were randomly collected from each treatment at 17:00 h (five eggs per replicate) on a weekly basis. The egg quality of the collected eggs was then determined at 20:00 h on the day of collection. The egg weight was measured using an egg multi tester (Touhoku Rhythm Co. Ltd., Tokyo, Japan). The eggshell breaking strength was evaluated using a model II egg shell force gauge (Robotmation Co., Ltd., Tokyo, Japan). Finally, a dial pipe gauge (Ozaki MFG. Co., Ltd., Japan) was used to measure the egg shell thickness, which was determined based on the average thickness of the rounded end, pointed end, and the middle of the egg, excluding the inner membrane. Digestibility of nutrients : After the conclusion of the feeding trial, six birds per treatment were randomly chosen for metabolic trials. The selected birds were individually housed in metabolic cages to determine the digestibility of nutrients. Laying hens were fed their respective diets containing chromic oxide ([Cr.sub.2][O.sub.3]) for 4 days prior to the collection period. All excreta of the birds were collected for 3 d. All the fecal samples along with feed samples, were then analysed according to the AOAC procedures (AOAC, 1990). The mineral contents were assayed using an inductively coupled serum emission spectrometre (Model JY-24, Jobin Yvon, Longjumeau, Cedex, France).

Serum calcium and phosphorus : At the beginning of the experiment, two birds per replicate were randomly selected and 5 ml of blood were collected from their left jugular veins using a sterilized injector. The samples were then transferred into a [K.sub.3]EDTA vacuum tube (Becton Dickinson Vacutainer Systems, Franklin Lakes, NJ, USA). At the end of the experiment, blood was collected from the same laying hens. The blood samples were then used to estimate the serum Ca (AOAC, 1990) and serum P concentrations (Fiske and Subba Row, 1925).

Statistical Analysis

All data were evaluated by analysis of variance following the GLM procedure in a completely randomised design. All analyses were conducted using the SAS software program (SAS Institute, 1996). Significant differences among the means of the treatment groups were determined at $p<0.05$ by Duncan's multiple-range test. Orthogonal comparisons were made using polynomial regression to measure the linear and quadratic effects of increasing dietary concentrations of supplemental phytase. According to the results of Duncan's multiple-range test, the effects of phytase supplementation of the NC diets at 250 FTU/kg were obviously lower than the effects of PC treatment; therefore, only differences between the PC group and the NC groups that received phytase supplementation at 500 FTU/kg were evaluated.

Results

Egg Production and Egg Quality

The results of the egg production and egg quality. No significant differences were observed in the feed intake among treatments. However, egg production, egg weight, egg shell breaking strength and egg shell thickness were significantly ($p<0.05$) reduced by 11.95%, 2.76%, 8.55%, and 3.47%, respectively, when laying hens fed the NC diet were compared to those that received the PC diet. Additionally, no linear effect in eggshell thickness was observed in response to phytase supplementation when compared to the NC diet, despite increases in thickness of 2.3% and 1.8% when hens were provided the NC diet supplemented with NAT and OPT at 500 FTU/kg, respectively. Conversely, as the level of phytase increased, increased egg production (Linear $p<0.05$) and eggshell breaking strength (Quadratic $p<0.05$) was observed. Moreover, a linear increase in egg weight ($p<0.05$) of 4.8% was observed in response to OPT supplementation, whereas no linear effect was observed in response to NAT supplementation. Furthermore, the effects of OPT-supplementation on egg production and egg weight were greater than those of NAT-supplementation ($p<0.05$).

Apparent Total Tract Nutrient Digestibility

The digestibilities are outline. The reduction of available P in the NC diet led to significant decreases in the digestibility of DM, N, Ca and P of 18.10%, 14.23%, 15.16% and 29.06%, respectively, when compared to the PC diet ($p<0.05$). However, the negative effects on digestibility were completely attenuated ($p<0.05$) by supplementation of the NC diet with 500 FTU/kg of phytase, with the exception of DM digestibility, which was significantly lower ($p<0.05$) in the NP2 and OP2 treatment groups compared to the PC treatment group. Furthermore, no significant differences were observed between groups supplemented with OPT and NAT.

Serum Calcium and Phosphorus

The data describing the serum Ca and P concentrations. The initial serum Ca and P levels are not presented in the table because of no significant difference in the serum Ca and P. Significant reductions of 16.7% and 38.5% were observed in the serum Ca and P concentrations, respectively, when birds that received the NC treatment were compared to those that received the PC treatment. In addition, the serum P concentrations increased linearly ($p<0.05$) with both forms of phytase when compared those observed in PC treatment. However, the linear effect ($p<0.05$) on serum Ca concentration was only observed in OPT group, whereas only a tendency ($p<0.10$) to increase was observed in the NAT group. No significant differences were observed between phytase sources. In addition, no significant differences were observed between the PC group and groups that received the NC diet supplemented with phytase at 500 FTU/kg. The beneficial effects of microbial phytase supplementation of P-deficient diets in poultry have been well documented (Adeola et al., 2004; Wu et al., 2006), which suggests that phytase supplementation can release phytate-bound nutrients and consequently improve nutrient utilisation.

The results of the present experiment appear to support this conclusion, as indicated by significant differences being observed between the PC and NC groups for most of the characteristics studied. These findings suggest that a diet containing 0.20% available phosphorus cannot meet the requirements of the 68-weeks laying hens. However, an experiment conducted by Augspurger et al. (2007) indicated that 0.20% available P is adequate for the older laying hens. This inconsistency suggests that the AP requirement for older laying hens should be revised.

However, our objective was to compare phytase sources, not determine or evaluate available phosphorus requirements. The differences between PC and NC that were observed in the current experiment allowed comparison of those two phytases. Most of the negative effects due to P-deficiency in the current experiment were almost attenuated by supplementation of the NC diet with phytase at 500 FTU/kg, which clearly demonstrates that both phytases used in this experiment have the ability to hydrolyse phytate-bound nutrients. Additionally, previous studies have revealed that negatively charged phytate can form insoluble complexes with positively charged proteins at low pH in the gastrointestinal tract (Cheryan, 1980; Reddy et al., 1982). Furthermore, it has been shown that phytic acid can form insoluble salts with minerals such as Ca, Mg, Fe, Zn and Cu (Bedford and Schulze, 1998; Liu et al., 1998), which can then restrict the utilisation of nutrient minerals and protein in nonruminant animals.

Comparison of Sources

Phytase derived from Escherichia coli (ECP) exhibited a broader range of optimal pHs (from 2.5 to 3.5), whereas phytase derived from Aspergillus niger (ANP) has a bimodal optimum pH of 2.5 and 5.5. Furthermore, ECP had 25% greater activity at pH 2.5 and 35% less activity at pH 5.5 when compared to ANP (Rodriguez et al., 1999a). A study conducted by Wodzinski and Ullah (1996) also reported that 6-phytases such as ECP would completely dephosphorylate the phytate molecule, whereas 3-phytases such as ANP do not due to their respective initiation sites. The differences in the biochemical and biophysical properties of phytase and the pH of the gut from which the phytate complex was liberated may lead to different levels of nutrients being released in response to different phytases.

However, in the current study, the only significant differences observed between groups treated with ECP (Optiphos) and ANP (Natuphos) were egg production and egg weight. Similar results were found by Stahl et al. (2000) and Sands et al. (2003), who reported that differences in biochemical and biophysical properties were not manifested with different phytases in young pigs and broilers, which was explained as the stomach of the animals used may not have proper pH to show the potential catalytic difference between these two phytase.

This may also explain why no significant differences were observed in most of the characteristics investigated in the current experiment. Additionally, previous studies conducted by Carlos and Edwards (1998), Ravindran et al. (1995) and Marounek et al. (2008) also suggested that laying hens utilise phytate P differently as they become older, and also suggested that the gastrointestinal pH or maturation of digestive tract would have changed as chicks become older, which consequently affect the potential phytase actively. Besides, researchers typically only measure phytase at pH 5.5 for dietary formulation, which may also lead to the different phytase activity observed in previous studies, as the pH of gastrointestinal tract from which the phytate complex was liberated is different. So it is appropriate to postulate that the equal effects of the both phytase observed in the current were the results of different gastrointestinal physiological in the older laying hen.

Feed Intake

In the current study, although the negative effects of providing 68-week-old laying hens P-deficient diets were quite evident, the severity of P deficiency was not manifested in terms of feed intake. This finding is similar to the results of Hughes et al. (2008), who reported that no significant effects on feed consumption were observed in response to either different NPP levels or supplemental phytaset of laying hens. However, Payne et al. (2005) and Wu et al. (2006) reported that feed intake increased linearly with the NPP levels in broiler and laying hens, respectively. And the negative effects were subsequently attenuated by phytase supplementation to the lower

NPP diets. Despite this finding, the exact reasons for the increase are not known; therefore, further studies should be conducted to evaluate the underline metabolism that how phytase supplementation affect the feed intake.

Egg Production

The egg production by laying hens decreased significantly by 11.95% in response to the NC diets. In addition, although supplementation of the NC diet with phytases led to a linear restoration of performance, the production of eggs by hens that received a diet supplemented with phytase at 500 FTU/kg was still significantly lower than that of hens that were fed the PC treatment. The results of the current experiment were partially consistent with those of studies conducted by Lim et al. (2003) and Wu et al. (2006), who found that supplementation of an NPP-deficient diet with phytase at 300 FTU/kg resulted in a significant improvement in egg production.

The results of the current study suggest that phytase supplementation at 500 FTU/kg induced a beneficial effect on egg production. This contention was supported by the observed improvement in digestibility and serum concentration in the blood, which are indicative of an increase in available nutrients. However, egg weight responded differently to the different phytases. Specifically, OPT supplementation led to a linear increase in egg weight until the levels observed in the PC treatment were attained, whereas no significant difference in the weight of eggs produced by hens that were fed the NAT diet was observed. Previous studies have also revealed controversial results regarding egg weight in response to phytase supplementation.

For example, Silversides et al. (2006) reported that phytase supplementation can enhance the egg production and egg weight of birds, whereas Wu et al. (2006) and Hughes et al. (2008) reported that phytase supplementation had no significant effect on egg weight. The variations in aforementioned studies may have attributed to the relationship between feed intake, egg production and egg weight, which is generally recognised that feed intake is positively associated with egg production while egg weight relates to egg production negatively. However, the reason may not explain the results in herein study, the results observed in current experiment may be due to the highly egg weight observed in the OPT2 treatment, which can attributed to the high phytase activity mostly attenuate the negative effect in the NC diets and partially postpone the depression of laying hen performance at this period. However, the exact reason for the results observed in the present experiment is unknown, further study may be conducted to evaluate the laying hen physiology and performance to phytase supplementation at this age.

Eggshell Quality

Eggshell quality is the most important egg quality to be considered in poultry breeding programs. Eggshell quality characteristics, such as shell thickness and shell breaking strength, primarily depend on Ca aggregation into calcium carbonate, organic materials, and trace minerals (Chowdhury and Smith, 2002; Mabe et al., 2003). In general, eggshell quality increases concurrently with the increase in digestibility that occurs in response to phytase supplementation. As expected, phytase supplementation exerted a beneficial effect on eggshell quality in the current experiment. Specifically, the eggshell breaking strength increased in a linear and quadratic manner in response to the addition of ECP and ADP, respectively. These findings are similar to the results of studies conducted by Um and Paik (1999) and Liu et al. (2007). However, no linear effect on eggshell thickness was observed in response to phytase supplementation, even though phytase supplementation at 500 FTU/kg led to an increase in the eggshell thickness to levels comparable to those of eggs produced by hens that received the PC diets, which indicate that supplemental phytase exerted a beneficial effect on eggshell thickness. The reason for the lack of a linear effect was not determined in this study; however, it may have occurred due to the small sample number, because of the large variation among individual eggs or as a result of replication and other systematic experimental errors.

Apparent Total Tract Nutrient Digestibility

The results of our metabolic trial indicated that the digestibility of P increased significantly in response to phytase supplementation when compared to the level observed in the NC group. These findings demonstrate that the phytate phosphorus was, to some extent, cleaved by phytase supplementation. Similar results were also observed by Liu et al. (2007), who reported a significant improvement in P digestibility in response to supplementation of the diet of Hy-Line brown layers with microbial phytase.

Additionally, the Ca and N digestibility was also increased with the supplementation of the phytase. At first glance, these results support that points that the phytate-bound minerals other than P were released in response to microbial phytase supplementation of the NC diets. However, based on the observation that the digestibility of Ca and N were also negatively impacted by reduced inclusion rate of inorganic phosphorus in NC diet compare to PC diet, it is presumptuous to state that any increase in digestibility of Ca and N due to phytase supplementation was due to a release of phytate-bound nutrients.

However, the effect of phytase on Ca and N digestibility remains unclear, it can be postulated that the previous increase may be more appropriately attributed to simply increased phosphorus absorption as a result of phytase, which is partially in agreement with Martinez-Amezcua et al. (2006), who suggested that it is possible to postulate that the positive effects of phytase on amino acid digestibility were due to alleviating the P deficiency, as Phosphorus is an important and necessary mineral for membrane function and active transporters such as the Na/K ATPase pump that are essential for amino acid absorption.

Similarly, to Ca digestibility, it is generally suggested that absorption and retention of Ca is influenced by the ratio of Ca: P, and study conducted by Underwood and Suttle (1999) also suggested that the improvement in skeletal Ca retention is accompanied by improved retention of P because Ca is only well utilised for skeletal growth when P is available at the same time, which may also supported the explanation that increased Ca digestibility is due to increased phosphorus absorption as a result of phytase Additionally, the Ca digestibility and N digestibility observed in this study were increased to levels observed in the PC treatment in response to phytase supplementation at 500 FTU/kg, whereas the digestibility of Ca, P and N was significantly lower in hens that received the NC diet supplemented with phytase at 250 FTU/kg than in those that received the PC diet. These findings indicate that phytase supplemental at 500 FTU/kg is sufficient to induce the hydrolysis of phytate in the 68-week-old of laying hens.

It is well known that the amount of dry matter (also known as dry weight) is a measurement of the mass of a sample that has been completely dried, and includes the weight of proteins, fat, milk, sugars and minerals in the sample. It is in the nature of things that phytase supplementation have positive effects on DM digestibility by inducing improved P, Ca and N digestibility. The results of previous studies have also suggested that phytase supplementation can improve DM digestibility by releasing bound organic nutrients such as protein and starch (Ravindran and Bryden, 1997). The significant difference in DM digestibility that was observed between hens that received the PC treatment and those that received phytase supplementation at 500 FTU/kg in the present study may be explained as follows: i) Other improvements in digestibility that were not investigated in the current experiment may have led to the significant effects observed in the factors evaluated here.

Serum Ca and P

Previous studies have stated that the serum P concentration seems to be less indicative of phytase efficacy than total tract P digestibility and retention of dietary P or the ash percentage of the 10th rib and metacarpals (Yi and Kornegay, 1996; Jongbloed and Mroz, 1999). However, in the current study, the increase in available minerals was also observed in the serum P and Ca of laying hens that were fed the phytase diet, and these values were found to be highly correlated with egg production and digestibility. Similar results were observed in a study conducted by Rama-Rao et al. (1999), who reported that the serum P increased linearly in response to phytase supplementation, but no linear effect was observed on serum Ca, and this may have occurred due to an antagonistic effect of serum Ca and P. In the current study, phytase supplementation led to an increase in serum Ca level to the level observed in the PC group.

The results of this study indicated that the negative effects of feeding hens 0.20% available phosphorus were quite evident for most of the characteristics that were investigated. However, supplementation of the diets with phytase at 500 FTU/kg can mostly compensate these negative effects in the current study. In addition, the abilities of NAT and OPT to liberate phytate-bound complexes were similar when they were included in 0.20% available phosphorus diets of laying hens. Taken together, these findings indicate that either source of phytase can be fed to commercial first cycle laying hens at 500 FTU/kg to effectively replace inorganic phosphorus when economically justified.

New Concepts on the Horizon: Phytogenics in Poultry Production

Modern poultry and egg production is facing several challenges. Growing demand for poultry products and rising prices for raw materials require the implementation of optimal production conditions with the aim to secure high animal performance. Phytogenic feed additives have gained considerable attention in the feed industry and producers are increasingly incorporating them into feeding programs.

Today, 61 (non-EU) or 70% (EU) of the companies are using phytogenic additives in broiler feeds (World Poultry, 2008). In comparison to Antibiotic Growth Promoters, phytogenics usually do not bear the risk of cross-resistances and residues in animal products. Improvements in feed conversion ratio (FCR) and body weight gain, as well as their benefits in helping to prevent intestinal diseases (such as Necrotic Enteritis) have been observed in recent trials with BIOMIN's phytogenic performance enhancer Biomin® P.E.P.

Information about the mode of action of commercially available phytogenic feed additives is rather scarce. It can be speculated that phytogenic feed additives vary greatly with regard to their *in vivo* effects in the animal. This variation is due to differences in the composition and biological activities of such feed additives. Assessment of biological effects is difficult if the composition of a test substance is unclear or variable. With Biomin® P.E.P. there is a defined phytogenic feed additive with standardised composition which is based on carefully selected raw materials.

What can we Expect from a Phytogenic Feed Additive?

Recent facts from studies with swine indicate that the mode of action of Biomin® P.E.P. is versatile and conclusive (Kroismayr, 2007). It was shown in this work that addition of Biomin® P.E.P. to basal diets resulted in a reduction of the total bacterial count in the intestine, increase of nutrient digestibility (Stony et al., 2006), decreased levels of microbial toxins in the gut and, therefore, a down-regulation of the immune system. Finally, this indicates that more energy and nutrients are available for accretion of body mass (energy and nutrient sparing). Furthermore, there is evidence that phytogenics stimulate digestive secretions such as saliva or endogenous digestive enzymes (Platel and Srinivasan, 1996).

Phytogenics in Broiler Production

Biomin® P.E.P. was tested in different dosages in a scientific trial at the Agricultural University of Athens, Greece. Dayold, male Cobb broiler chicks were assigned to different treatments, comprising 3 replications per treatment and 105 birds per treatment. The Negative Control (NC) contained no growth promoters, whereas the Positive Control (PC) contained Avilamycin. In further treatments, Biomin® P.E.P. 125 poultry was supplemented at 125 or 250 g/t, respectively. Biomin® P.E.P. increased body weight gain and significantly improved FCR. Differences between the dosages of Biomin® P.E.P. were minor, indicating that the regular inclusion level of 125 g/t was optimal under the experimental conditions herein. Similar positive results were obtained in a 45-day study conducted in Thailand. In this trial, the impact of Biomin® P.E.P. was investigated in one-day-old Ross broilers, which were assigned to 4 dietary treatments. The treatments were (1) Negative Control (NC), (2) Biomin® P.E.P., (3) Positive Control (PC) containing Flavomycin. Monensin was included in the starter and grower diets as anticoccidia agent. Compared to the NC, Biomin® P.E.P. increased daily weight gain by 4.2 %. FCR amounted to 1.77, 1.71 and 1.75 in the NC, Biomin® P.E.P. treatment and PC, respectively, and mortality was 1.12, 0.36 and 0.59%, respectively, in these treatments.

The Productivity Index [PI = Livability (%) x Live weight (kg)/ age (days)/FCR x 100] was 250, 255 and 235 in treatments 1, 2 and 3. Positive effects of phytogenics on broiler performance were also obtained by Cross et al. (2007) and Bölükbasi et al. (2006).

Phytogenics in Egg Production

The effects on performance and economics of Biomin® P.E.P. were investigated in the early stages of the egg production cycle in the Bangkok Animal Research Centre (BARC) (Nichol and Steiner, 2008). A 12-week trial was carried out with high-performing, female Lohmann Brown hens, using six replications with 16 birds per replication in a randomised complete block design and resulting in 96 hens per treatment. The age of the birds at the beginning and conclusion of the trial was 20 and 32 weeks, respectively. The birds were assigned to two dietary treatments: (1) Control (no additives), (2) Biomin® P.E.P. Hens fed phytogenics consumed less feed and had higher egg production as compared to the control birds. Total and average daily feed intake was lower by 1.8% when the control diet was supplemented with phytogenics.

Hens offered phytogenics produced more eggs and had a better feed conversion in comparison to birds in the control group. Additionally, supplementation of the diets with phytogenics improved egg shell parameters, i.e. shell thickness ($P<0.05$) and albumen height. As indicated by a higher Haugh Unit rating (82 vs. 79), the internal egg quality was higher in hens fed phytogenics.

Specific Applications of Phytogenics

As shown in the trial reports above, Biomin® P.E.P. works well as performance enhancer when it is included in the feed. Additionally, due to the effect of its active ingredients against several gram-positive and gram-negative bacteria, Biomin® P.E.P. represents a solution to prevent and overcome critical stages in the production cycle. Such stages include vaccination or a change of diets or feeding regimen, for example as the birds switch from starter to grower or from grower to finisher diets. Such changes disrupt the digestive process that can alter the balance of microorganisms in the gut. These disruptions invite pathogenic proliferation such as *Salmonella, Clostridia, E. coli,* and others. The use of Biomin® P.E.P. helps to prevent and overcome such challenges. In addition to the powdered formulations of Biomin® P.E.P. there is also the liquid formulation Biomin® P.E.P. sol, which is the ideal means for a short-term, high-dosage application to drinking water in these critical stages. Finally, recent work is in progress at the US Department of Agriculture (USDA) to identify the potential of phytogenics to reduce the clinical signs of Necrotic Enteritis (NE) in broilers. The first results are extremely promising, showing that Biomin® P.E.P. had a highly beneficial impact in alleviating NE symptoms in birds that had been challenged with Clostridium perfringens, which is the main causative agent of this damaging, costly disease (Mc Reynolds et al., 2008).

Phytogenics represent one of the most promising groups of performance-enhancing feed additives. It should be kept in mind that only a well-balanced and scientifically developed combination of active ingredients with defined properties can be expected to function synergistically in order to bring about the desired benefits for the producer. In this sense, Biomin® P.E.P. is the phytogenic solution of choice, combining a proven mode of action with consistent, positive effects on animal performance.

7

Chick Production in Broiler Breeders

There are several important instances in the management of broiler breeders where "more" is not "better". In simple terms, "more feed" allotted to hens to the extreme degree of ad libitum feeding reduces settable egg production by as much as 40 eggs per hen.

Breeder hens use dietary energy differently than egg-type hens. Young broiler breeder hen respond to extra dietary energy by developing extra ovarian follicles. Several research projects have been carried out at the University of Alberta, that have shown that "more ovarian development" does not mean "better reproductive efficiency". Clearly, a high priority during the early laying period is to promote ovarian development at a rate that will result in the highest number of settable eggs. This paper will look at the process of sexual maturation of the breeder pullet to improve egg production, fertility and hatchability.

The critical time in broiler breeder management is the period from photostimulation (lighting) to peak production. This period is characterised by relatively fast weight gains, and the changes brought about by the hormones being produced by a newly active ovary. Photostimulation is generally considered the cue to initiate puberty, although the response to light is readily modified by level of feeding.

Light energy passes through the skull and illuminates the hypothalamus. When the hypothalamus receives a photostimulatory signal (long day length above a threshold of intensity), hormones involved in ovarian function are produced. One of the first responses you can see by looking at the ovary of the bird after lighting is that the very small ovarian follicles begin to increase in size. The small follicles produce large quantities of estrogen, the hormone that causes most of the reproductive transformation associated with puberty. Firstly, estrogen influences the production of yolk precursors (building blocks of yolk production)in the liver of the bird. The liver visibly enlarges and becomes paler as its fat content increases for the production of egg yolk lipids.

Secondly, the oviduct increases in size so it is ready to receive ovulated follicles. This step appears to be a limiting one for heavy turkeys, as the ovary in these birds develops faster than the oviduct does, so there is a high incidence of follicles that are ovulated that are not picked up (internal ovulation) and hence, not formed into normal eggs.

Thirdly, estrogen influences changes to bone composition that allow calcium to be mobilised daily to make egg shells.

Finally, estrogen combines with male sex hormones (androgens) which results in changes to plumage, comb size and colour and sexual receptivity to males.

Shaver Starbro pullets were reared to 20 weeks of age to determine the influences of lighting programme and feed allocation from 20-25 weeks on ovarian development and production traits.

The experiment used two lighting programmes and two levels of feeding. A "slow photoperiod" treatment (SP) was gradually brought from 8L:16D with weekly increases in day length. A "fast photoperiod" treatment (FP) was increased from 8L:16D to 15L:9D in a single step at 20 weeks of age. The "slow feed" treatment group (SF) received very moderate increases in feed allocation from 20 to 25 wk of age. Increases in feed allocation of greater than 5 g, were divided up into two smaller increases per week.

The "fast feed" treatment group (FF) received a more generous feed increase between 20-25 weeks of age. It was not the original intent to decrease feed allocation to these birds; however, at 22 wk of age we discovered that the increases at 20 and 21 weeks of age had a greater effect on body weight that planned. The decision to decrease the feed allocation at 22 weeks of age was made to prevent the fast feed hens from becoming too heavy. It should be noted that the withdrawal of feed did not result in a decrease in body weight in subsequent weeks. From 25 to 64 weeks of age all of the birds were exposed to the same photoperiod (15L:9D) and a common feed allocation. A group of birds were killed on the day following sexual maturity to examine development of the reproductive tract and carcass composition.

Individual egg production records were kept and eggs were incubated at weekly intervals to assess treatment effects on chick production. Feed allocation was decreased after peak egg production to keep the birds at proper body weight targets (for age).

Body Weight: There were no significant differences in age at sexual maturity between any of the groups. Body weight and fat pad weights at sexual maturity were not different in terms of main effects. Within the group of slow photoperiod hens, the fast feed programme resulted in heavier body weights compared to the slow feed programme.

Ovary: Ovary morphology was highly influenced by both the photoperiod and feed allocation treatments. The slow photoperiod treatment had a greater ovary weight than did the fast photoperiod treatment. The fast feed treatment resulted in significantly heavier ovary weight and one more large follicle than the slow feed treatment. To our knowledge this is the first report of a relatively minor feeding difference affecting large follicle recruitment in commercial breeders.

Egg Production: All treatment groups had very high rates of total egg production when expressed in terms of main effects or the interactions of the main effects. There was a 10.9 egg advantage to use of the slow feed feeding programme compared to the fast feed programme. This fact is presumably related to improved control of the follicular recruitment process, as fast feed birds also had fewer large follicles at sexual maturity.

Fertility and Hatchability: Hatchability was reduced in the fast photoperiod programme compared to the slow photoperiod programme. Embryonic mortality was examined on day 7, 14, and 21 of incubation. The greatest embryonic mortality was found in the fast photoperiod treatment in the later stages of incubation.

Eggs from the fast fed hens had higher embryonic mortality than eggs from the slow fed hens. This indicates that developmental problems may be associated with the eggs from birds with excessive follicle development. These data suggest that minor differences in feed allocation have an influence on ovarian form and function. Proper management of feed allocation during the pre-lay phase is critical to Optimise the number of saleable chicks produced.

The above research data support the hypothesis that chick production in broiler breeder hens can be compromised by over-feeding early in lay. Future management programmes for breeder hens should focus on limiting excessive follicle production so that there is an orderly cooperation between the ovary and the oviduct in the production of hatching eggs. Such a programme would minimise egg shell problems, reduce the incidence of multiple-yolked eggs and result in long egg-laying sequences. Long egg-laying sequences lead to high peak production and excellent persistency of lay. Further studies on limiting follicular growth by controlling energy intake and where nutrients are partitioned or distributed through the body and the role of light in sexual maturation are underway at the University of Alberta.

Broiler

A broiler is a type of chicken raised specifically for meat production. Modern commercial broilers, typically known as Cornish crosses or Cornish-Rocks are specially bred for large scale, efficient meat production and grow much faster than egg or traditional dual purpose breeds. They are noted for having very fast growth rates, a high feed conversion ratio, and low levels of activity. Broilers often reach a harvest weight of 4-5 pounds dressed in only five weeks.

They have white feathers and yellowish skin. This cross is also favourable for meat production because it lacks the typical "hair" which many breeds have that necessitates singeing after plucking. Both male and female broilers are slaughtered for their meat. In 2003, approximately 42 billion broilers were produced, 80% of which were produced by four companies: Aviagen, Cobb-Vantress, Hubbard Farms, and Hybro.

History

Before the development of modern commercial meat breeds (cows, chickens, etc.) broilers consisted mostly of young male chickens (cockerels) which were culled from farm flocks. The males were slaughtered for meat and the females (pullets) were kept for egg production. Compared to today, this made chicken meat scarce and expensive compared to eggs, and chicken was a luxury meat. The development of special broiler breeds decoupled the supply of broilers from the demand for eggs.

This, along with advances in nutrition and incubation that allowed broilers to be raised year-round, allowed chicken to become a low-cost meat. Broilers are often called "Rock-Cornish," referring to the adoption of a hybrid variety of chicken produced from a cross of male of a naturally double breasted Cornish strain and a female of a tall, large boned strain of white Plymouth Rocks. This first attempt at a hybrid meat breed was introduced in the 1930s and became dominant in the 1960s. The original cross was plagued by problems of low fertility, slow growth, and disease susceptibility, and modern broilers have gradually become very different from the Cornish x Rock hybrid.

Modern Variants

Access to a special diet of high protein feed delivered via an automated feeding system. This is combined with artificial lighting conditions to stimulate growth and thus the desired body weight is achieved in 4 - 8 weeks, depending on the approximate body weight required by the processing plant. After processing, the poultry is delivered as fresh or frozen chicken to the stores and supermarkets.

Figure: *Five day old broiler strain Cornish-Rock chicks*

Because of their efficient meat conversion, broiler chickens are also popular in small family farms in rural communities, where a family will raise a small flock of broilers. Broilers are sometimes reared on a grass range using a method called pastured poultry, as developed by Joel Salatin and promoted by the American Pastured Poultry Producers Association. The term "broiler" is widely known in North America, Australia and England but not elsewhere in the English speaking world. The term "broiler chicken" is very widely used in Pakistan and India, as it was in the former German Democratic Republic and still nowadays in some eastern parts of Germany. The term is also used in Bangladesh, Indonesia, Sweden, Nigeria, Finland, Poland, Turkey and the Balkans.

Broiler Health Issues

Broiler chickens may develop several health issues as a result of selective breeding. Broiler chickens are bred to be very large to produce the most meat per animal. The large chickens cannot stand because their bodies grow too quickly for their legs. Therefore, they may become lame or suffer from broken legs. Broiler chickens are also prone to heart attacks for the same reason, as the heart cannot support blood flow to the large body of the chicken. Another issue with selective breeding is the larger chickens have a more aggressive appetite. The broilers are feed restricted and this leads to behavioural issues in chronically hungry birds.

Broiler chickens may often get joint disorders because their legs cannot bear the heavy bodies. A Swedish study by SLU Skara (Swedish farming university) revealed that only 1/3 of studied broiler chickens that were about to be slaughtered were healthy. Additionally, it is very inactive and as a result is a poor forager, prone to predation, and is generally not suited to small free range homestead flocks.

If the litter in the pen is not properly managed to prevent birds from standing and resting in their feces, painful hock burns and foot ulcerations and blisters can occur. Pastured birds which are rotated frequently typically do not have these issues.

Leptin Receptor in the Chicken Ovary: Potential Involvement in Ovarian Dysfunction of Ad Libitum-fed Broiler Breeder Hens

The ovary of the mature hen contains a hierarchy of yellow yolky follicles and several thousand smaller follicles from which the large yolky follicles are recruited. The yellow follicles are arranged in a size hierarchy and are committed to ovulation. In each follicle, the granulosa cells are surrounded by theca tissue and are separated from it by a basement membrane. Each compartment of the largest yellow follicles (theca and granulosa cells) can be anatomically separated to follow the individual functions of these two ovarian compartments in follicle growth and differentiation.

At an early stage of follicular development the small ovarian follicles produce estrogens and androgens. As follicles begin to sequester yolk their production of estrogens from theca cells decreases to become very low at ovulation. As follicles are recruited into the yolky follicular hierarchy, estrogen and androgen production by theca cells diminishes and production of progesterone by granulosa cells increases. The largest F1 yellow follicle then attains the highest progesterone production at the time of ovulation. In most domestic animals, reproductive function is considerably affected by nutrition. Various hormones, including growth hormone, insulin-like growth factors (IGFs) and insulin, have been proposed as potential mediators affecting reproductive function. However, the interactions between the reproductive endocrine axis and the metabolic axis have not been clearly determined. Leptin represents also a good candidate for such reproductive-metabolic interactions. Leptin, the protein hormone synthesised and secreted mainly by adipose tissue, has primarily been shown to regulate food intake and energy expenditure. Recent studies have demonstrated that leptin may also be involved in the regulation of reproductive mechanisms in human and rat ovaries.

Exogenous leptin can rescue reproductive function in ob/ob leptin-deficient mice that are not only obese but also infertile. This leptin action is independent from weight loss since feed restriction in ob/ob female mice fails to restore fertility. Leptin can also advance the onset of puberty or at least reverse the delay caused by feed restriction in rodents. In chickens, leptin attenuates the negative effects of fasting on ovarian function.

Injections of leptin during fasting delays cessation of egg laying, attenuates regression of yellow hierarchical follicles, alters ovarian steroidogenesis and limits apoptosis. Leptin exerts its effect by binding to a receptor which belongs to the cytokine receptor super-family. The chicken leptin receptor has been cloned and sequenced. Its expression at the level of the ovary suggests that leptin might act directly on the ovary to regulate chicken reproductive function.

Standard broiler breeders have been submitted to high selection pressure on growth and feed efficiency. Male traits have been favoured resulting in poorer reproductive performances of the hens. As a result of such selection, broiler breeder hens are subject to metabolic disorders and reproductive dysfunction. Overfeeding during reproductive development is associated with the formation of excessive numbers of ovarian yellow follicles which can be arranged in multiple hierarchies, with increased production of unsettable eggs. Severe feed restriction during rearing reduces the production of yellow follicles, the incidence of double ovulation and considerably improves the laying rate. Up-regulation of yellow follicles has been related to excessive recruitment and rapid growth rate of follicles to maturity, especially under ad libitum feeding. However, the mechanisms that regulate these processes are still not fully explained.

This study investigated the potential involvement of leptin and its receptor in ovarian abnormalities observed in broiler breeder hens fed ad libitum. We aimed to determine the evolution of leptin receptor gene expression in both granulosa and theca cells from the four largest yellow follicles of 32 week-old hens during the laying period. The effects of genotype and diet on plasma leptin levels and ovarian expression of the leptin receptor gene were also quantified.

For this purpose standard broiler breeder hens fed ad libitum or feed-restricted were compared to a French "Label" genotype and a dwarf "Experimental" line. Compared to fast growing standard broiler breeders, the French "Label" is a dwarf, slow-growing broiler genotype and the "Experimental" line is a dwarf genotype with a growth potential of progeny chicks close to that of the standard broiler chicks. The "Experimental" line is specifically selected for reproductive traits and viability at partial expenses of growth performances. The "Label" line tolerates ad libitum feeding of breeders and does not have reproductive problems under ad libitum feeding whereas the «Experimental» line can be fed ad libitum on a low energy diet during the growing period but presents more reproductive problems than the Label genotype but less than the standard genotype fed ad libitum.

Methods

Animals: Three lines of broiler breeder hens supplied by Hubbard primary breeder (Chateaubourg, France) were used in this experiment. The S line is a standard fast growing broiler, the French Label (L) line is a slow growing broiler breeder strain used for the quality market and the experimental (E) line is a broiler breeder strain bearing the "dw" dwarf gene.

This strain has a decreased need for rationing. S and L hens were given the same regime in accordance with Hubbard nutritional recommendations (2724 kcal/kg) in fine meal form. During the growing period (0–20 weeks), half of the S hens were feed restricted (SR) on the same diet in order to match a reference body weight curve provided par Hubbard primary breeder, the other half (SA) and all the L hens were fed ad libitum.

The feed intake was equivalent to 37% of the SA group up to point of lay. A special diet was designed for the E hens, consisting of a series of finely ground meal diets with a lower energy content (2550 kcal/kg). The interest of the E group is mainly practical, it represents an actual alternative for severely restricted standard broiler breeder hens. Transition between the grower and breeder feed occurred at 20 weeks of age for the ad libitum-fed hens (L, SA and E). Transition occurred at the beginning of laying for the restricted hens that were then allowed ad libitum access to food.

At 32 weeks of age, blood samples were collected from 12 hens of each experimental group (SA, SR, E and L) and six hens were sacrificed by an overdose of pentobarbital (Sanofi-Sante Animale, Libourne, France). The ovaries and liver were immediately removed. Granula and theca compartments from the first (F1), second (F2), third (F3) and fourth (F4) largest ovarian yellow follicles were dissected as previously described.

Since SA birds presented a greater average number of yellow follicles per ovary (9.36, 8,00, 7.42, and 6.33 yellow follicles/ovary for the SA, SR, E and L hens respectively) and a higher proportion of pairs of yellow ovarian follicles undergoing simultaneous development, follicles were assigned to the same follicular rank when the difference of weight between two follicles was of 0.4 g or less. In that case, one follicle per pair was collected and dissected. Tissues were immediately snap frozen in liquid nitrogen and stored at -80°C until used for total RNA extraction. This experiment was carried out with due regard to the legislation governing ethical treatment of animals, and investigators were certificated by the French government to carry out animal experiments.

RNA Extraction and Leptin Receptor RT-PCR

Total RNA was extracted from liver, granulosa and theca cells using RNA InstaPure (Eurogentec, Angers, France) according to the manufacturer's recommendations. After DNAse treatment using Ambion's DNA-free kit (Clinisciences, Montrouge, France), 2 µg of total RNA were reverse-transcribed (RT) in a final volume of 20 µl using RNAse H^- MMLV reverse transcriptase (Superscript II, Invitrogen, Cergy Pontoise, France) and random hexamer primers (Promega, Charbonnières, France). cDNA was then diluted to 1:8.

For normal PCR amplification, five microliters of the RT reaction were amplified for 35 cycles in a 50 µl reaction volume containing 2.5 units of Taq DNA Polymerase (Amersham Biosciences, Orsay, France), 2.5 mM $MgCl_2$, 0.2 mM dNTPs (Promega, Charbonnieres, France) and 0.2 µM of each forward and reverse primer. Leptin receptor forward (5'-GTC CAC GAG ATT CAT CCC AG-3') and reverse (5'-CCT GAG ATG CAG AGA TGC TC-3') primers were chosen according to the previously determined sequence of the chicken leptin receptor cDNA.

This pair of primers amplifies a 271 bp cDNA fragment located in the coding sequence of the extra-cellular domain. The amplification conditions were as follows: denaturation at 94°C for 30 sec, annealing at 60°C for 30 sec and primer extension at 72°C for 60 sec. After final extension at 72°C for 10 min, PCR products were resolved on 1.5% agarose gel containing ethidium bromide.

Real-Time RT-PCR

Real-time RT-PCR was performed as previously described. Briefly, forward leptin receptor primer 5'-GCATCTCTGCATCTCAGGAAAGA-3' and reverse leptin receptor primer 5'-GCAGGCTACAAA CTAACAAATCCA-3'(nucleotides 362 to 448 of the chicken leptin receptor cDNA sequence) were designed to be intron-spanning to avoid co-amplification of genomic DNA using Primer Express Software (Applied Biosystems, Courtaboeuf, France). A 20 μl master mix containing 12.5 μl SYBR Green PCR Master Mix, 1 μl forward primer (300 nM), 1 μl reverse primer (300 nM) and 5.5 μl water was prepared to perform real-time PCR (Applied Biosystems, Courtaboeuf, France). Five microliters of cDNA dilution was added to the PCR Master Mix to a final volume of 25 μl. The following PCR protocol was used on the ABI Prism 7000 apparatus (Applied Biosystems, Courtaboeuf, France): initial denaturation (10 min at 95°C), followed by a two-step amplification programme (15 sec at 95°C, followed by 1 min at 60°C) repeated 40 times. Quantification was performed using ABI integrated software as previously described. 18S ribosomal RNA was chosen as the reference gene. The level of 18S RNA was determined using the Pre-developed TaqMan Ribosomal RNA control kit (Applied Biosystems, Courtaboeuf, France) according to the manufacturer's recommendations. The results were expressed as the leptin receptor mRNA/18S RNA ratio. Each PCR run included a no template control and replicates of control and unknown samples. Runs were performed in triplicate.

Plasma Lipid, Glucose and Hormone Analyses

Total cholesterol, phospholipid, and triglyceride plasma concentrations were calculated with "Cholesterol RTU", "Phospholipides Enzymatique PAP 150", and "Triglycerides Enzymatique PAP 150" kits (bioMérieux, Charbonnieres les Bains, France) according to the manufacturer's recommendations. Plasma glucose levels were measured by the glucose oxidase method using an automated analyser. Plasma insulin levels were determined by a radioimmunoassay with a guinea pig anti-porcine insulin antibody using chicken insulin as the standard. Plasma concentrations of leptin were determined by a multi-species leptin RIA kit (LINCO Research Inc, CliniSciences, Montrouge, France) according to the recommendations of the manufacturer.

Statistical Analysis

The results of plasma lipid, glucose and hormone analyses as well as leptin receptor mRNA expression in the liver were analysed by one-way ANOVA and means were compared by Student Newman Keuls multiple comparison test. The effects of the groups of birds (E, L SA and SR), follicular rank and possible interaction on the logarithm of leptin receptor mRNA levels were tested by two-way ANOVA using the General Linear Model (GLM) procedure of SAS (SAS Institute, 1999. SAS User' Guide, Version 8 ed. SAS Institute Inc., Cary, NC). An additional effect of the subject was introduced into the model in order to take into account the fact that measurements of leptin receptor mRNA expression for the different follicular ranks were performed on the same animal. Pairwise comparisons of means for each significant effect of the ANOVA were performed by Scheffe test with the least means squares statement of the GLM procedure. The level of significance was set at $P < 0.05$.

Results

Leptin Receptor MRNA Expression in Granulosa and Theca Cells: We demonstrated the expression of leptin receptor mRNA in granulosa and theca cells from the three genotypes fed ad libitum or restricted for the S line. The expression of leptin receptor mRNA for both ovarian cells was detected in each hierarchical yellow follicle studied (F1 to F4).

Evolution of Leptin Receptor MRNA Expression with Follicular Development

The evolution of expression of leptin receptor mRNA during follicle development was investigated in both granulosa and theca cells from F4 to F1 yellow follicles using real-time RT-PCR. Since leptin receptor mRNA levels did not follow a normal distribution (skewness of 5.48 and Kurtosis of 32.88), they were log transformed. The resulting distribution was closer to the normality with skewness of 0.87 and kurtosis of 0.69. Variance analysis was performed on transformed data.

In the E line, expression of the leptin receptor mRNA decreased between F4 and F1 yellow follicles. However the high variability of the expression of leptin receptor mRNA in F4 follicles prevented the decrease from reaching statistical significance. Compared to the L line, the level of expression of the leptin receptor in the granulosa cells was lower in the E group, especially for the F4 and F3 yellow follicles but statistical significance was reached only for the F4 follicles. In the S line fed ad libitum, expression of the leptin receptor in the granulosa was dramatically up-regulated. This up-regulation was clearly evident in F4 F3 and F1 follicles. Wide variability was also observed in F4 and F3 follicles. Feed restriction of the standard hens (SR) induced a general decrease in the expression of leptin receptor mRNA. The overall level of expression of the leptin receptor mRNA measured in the SR line was similar to that observed in the L and E lines. Compared to the SA birds, feed restriction has restored the decrease in the expression of leptin receptor mRNA with follicle development.

Expression of leptin receptor mRNA in the theca cells remained stable during yellow follicle development, whatever the group of birds considered. Statistical analysis did not reveal any difference in leptin receptor mRNA expression between the 4 groups of birds.

Expression of Leptin Receptor MRNA in the Liver

In the liver, the expression of leptin receptor mRNA was up-regulated in S birds fed ad libitum. Feed restriction of S birds restored the level of expression of leptin receptor mRNA similar to that measured in the E and L birds.

Plasma Lipid and Glucose and Hormone Concentrations

At 32 weeks of age plasma leptin and insulin concentrations were found to be similar in the three genotypes. Food restriction of the standard hens did not alter plasma leptin, glucose or insulin levels. Triglyceride, cholesterol and phospholipid levels were also measured. Cholesterol and phospholipid levels were not affected by genotype or diet. On the other hand, triglyceride levels seemed to be affected by the genotype. Lower triglycerides levels were measured in Standard birds (SA, SR). However, statistical significance was reached only for the restricted SR birds.

Discussion

Several studies conducted on theca and granulosa cells have shown that leptin may have direct negative effects on ovarian steroidogenesis in various mammalian species. Leptin inhibits insulin-induced progesterone and 17β-estradiol production by isolated bovine granulosa cells and impairs the hormonally-stimulated in vitro release of 17β-estradiol by rat granulosa cells. In granulosa cells from fertile women, leptin inhibits FSH and IGF-I stimulated estradiol production

Since leptin has a more potent inhibitory action of insulin-induced aromatase activity of granulosa cells from small than large follicles, it has been proposed that the numbers of leptin receptors in granulosa cells might decrease as follicles develop in order to make mature Graafian follicles less sensitive to the negative action of leptin. As shown in this study and in previous reports, the leptin receptor was expressed in the hen ovary in both granulosa and theca cells, suggesting a direct action of leptin at the level of the ovary.

It seemed that leptin might affect ovarian steroidogenesis in laying hens during fasting but the involvement of leptin on steroidogenesis during normal follicle development remained to be determined. In this study we demonstrated that the direct action of leptin on the ovary might be modified during follicle development since the level of expression of its receptor clearly decreased during maturation of yellow follicles. This decrease was particularly evident in slow growing broiler breeder hens from the "Label" genotype and from the feed-restricted standard line. Given that fast growing chickens (ad libitum-fed standard and Experimental broiler breeder hens) have the highest reproductive problems, genetic or nutritional control of the growth rate might regulate ovarian leptin receptor gene expression and improve reproductive function. Such evolution of receptor expression in the follicular hierarchy has previously been shown for the FSH receptor. FSH-stimulated steroidogenesis declined during follicle maturation and was associated with a decrease in FSH receptor numbers. Conversely, the expression of mRNA encoding the IGF-I receptor and the related efficacy of binding of IGF-I to granulosa cells increased as the follicle matured. Since leptin receptor gene expression was modified during follicle development, leptin might also be involved in regulation of the follicular hierarchy and onset of pre-ovulatory steroidogenesis, as has been proposed for gonadotrophins and growth factors including FSH and IGF-I.

Unlike mammals, progesterone in chickens is synthesised and secreted mainly by granulosa cells whereas theca cells generate estradiol. Progesterone produced by granulosa cells from mature follicles provides the positive feedback necessary to stimulate a pre-ovulatory surge of LH. IGF-I has been involved in the regulation of ovarian steroidogenesis in both mammals and birds.

IGF-I stimulates progesterone production from avian granulosa cells whereas it up-regulates estradiol from mammalian granulosa cells. Since leptin is considered to be an inhibitor of insulin and IGF-

I action on steroidogenesis in mammals, leptin might have similar negative action in birds. Thus, the decrease in its receptor in the granulosa suggests that the inhibiting action of leptin would decrease during follicle development and consequently favours the stimulatory effect of gonadotrophins and IGF-I on follicular maturation.

This hypothesis is also consistent with the weaker steroidogenic response of granulosa cell culture of ad-libitum fed standard broiler breeder hens when stimulated by IGF-I compared to granulosa cell culture from feed-restricted birds. Moreover, Onagbesan et al (2004) demonstrated in an experiment similar to that performed in the present study and using the same genotypes that plasma progesterone levels were clearly affected in SA birds.

They demonstrated that plasma progesterone levels remained relatively stable between 25 and 37 weeks of age in the E, L and SA birds with a significant lower level in the SA birds (2.2 ± 0.62 ng/ml for SA birds compared to 3.9 ± 0.36 and 4.2 ± 0.54 for L and E birds respectively). In restricted standard birds, plasma progesterone levels dramatically increased and reached values (3.8 ± 0.26 ng/ml) closed to that measured in the E and L lines (Onagbesan et al., 2004, data from progesterone levels were personal communication from Dr Onagbesan, Catholic University of Leuven, Belgium).

The erratic pattern of oviposition in standard broiler breeder hens fed ad libitum has been previously demonstrated to be related to abnormal maturation of steroidogenesis, particularly in the two largest yellow follicles. Since F2 and F1 yellow follicles presented similar endocrine profiles, the pre-ovulatory surge of LH probably triggers ovulation of the two largest follicles. In this study we have shown that ad libitum feeding of broiler breeder hens dramatically up-regulated expression of the leptin receptor in the granulosa cells of yellow follicles and changed the evolution of expression of this receptor with follicle development. These results suggest a strong action of leptin on the ovaries of ad libitum fed birds.

Feed restriction reduced the level of expression of the leptin receptor and on the whole restored the evolution of expression of the receptor with follicle maturation. Since ad libitum feeding affects the hierarchical endocrine order of the follicles, as a potential inhibitor of hormonally induced avian steroidogenesis leptin represented a good candidate to explain the affects of follicular hierarchy. Up-regulation of the expression of the leptin receptor gene was also demonstrated in the liver. This up-regulation may be related to the control of lipogenesis. The liver plays a key role in lipid metabolism and lipogenesis in avian species and the standard broiler breeder hens were the fattest birds of this experiment.

Since expression of its receptor was dramatically up-regulated in SA hens, leptin probably played an important role in the increased number of large yellow follicles and abnormal follicle hierarchy. However the factors involved in regulation of the expression of the leptin receptor within the hen ovary remains to be determined.

Among the plasma hormones and lipids analysed in this study only triglycerides were found to be different between strains, with a lower level in the restricted standard broiler breeder hens that were also the leanest birds.

Down regulation of the expression of the leptin receptor by homologous and heterologous signals have previously been demonstrated in both mammals and chickens.

Leptin and insulin are able to down-regulate expression of the chicken leptin receptor in vitro. In the present study, plasma leptin and insulin levels were similar for each genotype and were not altered by feed restriction in the standard genotype. We have previously demonstrated that during the first 5 weeks of age, plasma leptin levels remained relatively stable in both broiler and layer chicken despite increased body weight.

However the absence of leptin levels differences may be related to the fact plasma leptin levels were measured at 32 weeks of age. SR birds were relaxed at the start of lay, switched to breeding feeding and allowed ad libitum access as the other groups of birds. We therefore suggested that leptin and insulin are probably not involved in the regulation of ovarian leptin receptor gene expression in ad libitum or feed-restricted standard broiler breeder hens.

Evidence of the regulation of expression of the leptin receptor gene in the granulosa related to follicle maturation and nutritional state strongly suggest that leptin played an important local and sequential role in the dysfunction of the follicular hierarchy observed in standard broiler breeder hens fed ad libitum.

This study suggests that the level of expression of the leptin receptor regulates the action of its ligand at the level of the ovary. This provides an interesting perspective to understanding the physiological role of leptin in the ovary.

Chicken

The chicken (*Gallus gallus domesticus*) is a domesticated fowl, a subspecies of the Red Junglefowl. As one of the most common and widespread domestic animals, and with a population of more than 24 billion in 2003, there are more chickens in the world than any other species of bird. Humans keep chickens primarily as a source of food, consuming both their meat and their eggs.

The traditional poultry farming view of the domestication of the chicken is stated in *Encyclopaedia Britannica* (2007): "Humans first domesticated chickens of Indian origin for the purpose of cockfighting in Asia, Africa, and Europe.

Very little formal attention was given to egg or meat production..." Recent genetic studies have pointed to multiple maternal origins in Southeast, East, and South Asia, but with the clade found in the Americas, Europe, the Middle East and Africa originating in the Indian subcontinent.

From India the domesticated fowl made its way to the Persianised kingdom of Lydia in western Asia Minor, and domestic fowl were imported to Greece by the fifth century BC. Fowl had been known in Egypt since the 18th Dynasty, with the "bird that lays every day" having come to Egypt from the land between Syria and Shinar, Babylonia, according to the annals of Tutmose III.

Terminology

In the UK, Ireland and Australia adult male chickens over the age of 12 months are primarily known as *cocks*, whereas in America and Canada they are more commonly called *roosters*. Males under a year old are *cockerels*. Castrated roosters are called *capons* (surgical and chemical castration are now illegal in some parts of the world). Females over a year old are known as *hens*, and younger females are *pullets*. In Australia and New Zealand (also sometimes in Britain), there is a generic term *chook* to describe all ages and both sexes. Babies are called *chicks*, and the meat is called *chicken*.

"Chicken" originally referred to chicks, not the species itself. The species as a whole was then called *domestic fowl*, or just *fowl*. This use of "chicken" survives in the phrase "Hen and Chickens", sometimes used as a British public house or theatre name, and to name groups of one large and many small rocks or islands in the sea. In the Deep South of the United States chickens are also referred to by the slang term *yardbird*.

General Biology and Habitat

Chickens are omnivores. In the wild, they often scratch at the soil to search for seeds, insects and even larger animals such as lizards or young mice.

Figure: *The adult rooster can be distinguished from the hen by his larger comb*

Roosters can usually be differentiated from hens by their striking plumage of long flowing tails and shiny, pointed feathers on their necks (*hackles*) and backs (*saddle*) which are typically of brighter, bolder colours than those of females of the same species. However, in some breeds, such as the Sebright, the rooster has only slightly pointed neck feathers, the same colour as the hen's.

The identification must be made by looking at the comb, or eventually from the development of spurs on the male's legs (in a few breeds and in certain hybrids the male and female chicks may be differentiated by colour). Adult chickens have a fleshy crest on their heads called a *comb or cockscomb*, and hanging flaps of skin either side under their beaks called *wattles*. Both the adult male and female have wattles and combs, but in most breeds these are more prominent in males. A *muff* or *beard* is a mutation found in several chicken breeds which causes extra feathering under the chicken's face, giving the appearance of a beard.

Domestic chickens are not capable of long distance flight, although lighter birds are generally capable of flying for short distances, such as over fences or into trees (where they would naturally roost). Chickens may occasionally fly briefly to explore their surroundings, but generally do so only to flee perceived danger. Chickens are gregarious birds and live together in flocks. They have a communal approach to the incubation of eggs and raising of young. Individual chickens in a flock will dominate others, establishing a "pecking order", with dominant individuals having priority for food access and nesting locations. Removing hens or roosters from a flock causes a temporary disruption to this social order until a new pecking order is established. Adding hens—especially younger birds—to an existing flock can lead to violence and injury.

Hens will try to lay in nests that already contain eggs, and have been known to move eggs from neighbouring nests into their own. Some farmers use fake eggs made from plastic or stone (or golf balls) to encourage hens to lay in a particular location.

The result of this behaviour is that a flock will use only a few preferred locations, rather than having a different nest for every bird. Hens can also be extremely stubborn about always laying in the same location.

It is not unknown for two (or more) hens to try to share the same nest at the same time. If the nest is small, or one of the hens is particularly determined, this may result in chickens trying to lay on top of each other.

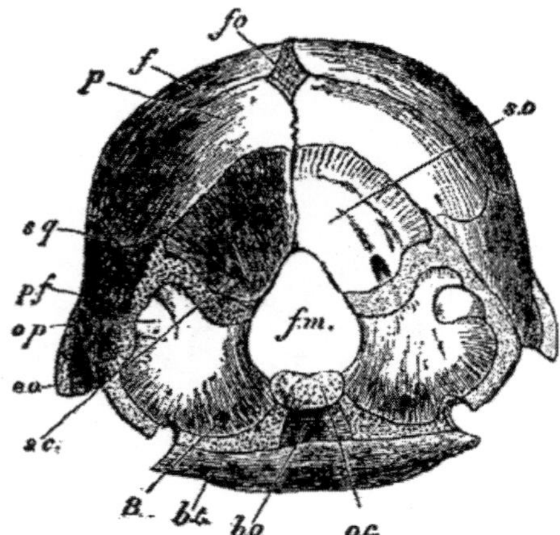

Figure: *Skull of a chicken three weeks old. Here the opisthotic bone appears in the occipital region, as in the adult Chelonian. bo = Basi-occipital, bt = Basi-temporal, eo = Opisthotic, f = Frontal, fm = Foramen magnum, fo = Fontanella, oc = Occipital condyle, op = Opisthotic, p = Parietal, pf = Post-frontal, sc = Sinus canal in supra-occipital, so = Supra-occpital, sq = Squamosal, 8 = Exit of vagus nerve.*

Roosters crowing (a loud and sometimes shrill call) is a territorial signal to other roosters. However, crowing may also result from sudden disturbances within their surroundings. Hens cluck loudly after laying an egg, and also to call their chicks. Chickens also give a low "warning call" when they think they see a predator approaching.

In 2006, scientists researching the ancestry of birds "turned on" a chicken recessive gene, *talpid2*, and found that the embryo jaws initiated formation of teeth, like those found in ancient bird fossils. John Fallon, the overseer of the project, stated that chickens have "...retained the ability to make teeth, under certain conditions...."

Food Sharing and Courting

When a rooster finds food, he may call other chickens to eat first. He does this by clucking in a high pitch as well as picking up and dropping the food. This behaviour may also be observed in mother hens to call their chicks and encourage them to eat.

To initiate courting, some roosters may dance in a circle around or near a hen ("a circle dance"), often lowering his wing which is closest to the hen. The dance triggers a response in the hen's brain, and when the hen responds to his "call", the rooster may mount the hen and proceed with the fertilization.

Breeding

Origins

The domestic chicken is descended primarily from the Red Junglefowl (*Gallus gallus*) and is scientifically classified as the same species. As such it can and does freely interbreed with populations of red jungle fowl. Recent genetic analysis has revealed that at least the gene for yellow skin was incorporated into domestic birds through hybridisation with the Grey Junglefowl (*G. sonneratii*).

The traditional poultry farming view is stated in *Encyclopaedia Britannica* (2007): "Humans first domesticated chickens of Indian origin for the purpose of cockfighting in Asia, Africa, and Europe. Very little formal attention was given to egg or meat production... " In the last decade there have been a number of genetic studies. According to one study, a single domestication event occurring in the region of modern Thailand created the modern chicken with minor transitions separating the modern breeds. However, that study was later found to be based on incomplete data, and recent studies point to multiple maternal origins, with the clade found in the Americas, Europe,

Middle East, and Africa, originating from the Indian subcontinent, where a large number of unique haplotypes occur. It has been claimed (based on paleoclimatic assumptions) that chickens were domesticated in Southern China in 6000 BC. However, according to a recent study, "it is not known whether these birds made much contribution to the modern domestic fowl. Chickens from the Harappan culture of the Indus Valley (2500-2100 BC), in what today is Pakistan, may have been the main source of diffusion throughout the world."

A northern road spread chicken to the Tarim basin of central Asia, modern day Iran. The chicken reached Europe (Romania, Turkey, Greece, Ukraine) about 3000 BC. Introduction into Western Europe came far later, about the 1st millennium BC. Phoenicians spread chickens along the Mediterranean coasts, to Iberia. Breeding increased under the Roman Empire, and was reduced in the Middle Ages. Middle East traces of chicken go back to a little earlier than 2000 BC, in Syria; chicken went southward only in the 1st millennium BC. The chicken reached Egypt for purposes of cock fighting about 1400 BC, and became widely bred only in Ptolemaic Egypt (about 300 BC). Little is known about the chicken's introduction into Africa. Three possible ways of introduction in about the early first millennium AD could have been through the Egyptian Nile Valley, the East Africa Roman-Greek or Indian trade, or from Carthage and the Berbers, across the Sahara.

The earliest known remains are from Mali, Nubia, East Coast, and South Africa and date back to the middle of the first millennium AD. Domestic chicken in the Americas before Western conquest is still an ongoing discussion, but blue-egged chicken, found only in the Americas and Asia, suggest an Asian origin for early American chickens.

A lack of data from Thailand, Russia, the Indian subcontinent, Southeast Asia and Sub-Saharan Africa makes it difficult to lay out a clear map of the spread of chickens in these areas; better description and genetic analysis of local breeds threatened by extinction may also help with research into this area.

Current

Under natural conditions, most birds lay only until a clutch is complete, and they will then incubate all the eggs. Many domestic hens will also do this–and are then said to "go broody". The broody hen will stop laying and instead will focus on the incubation of the eggs (a full clutch is usually about 12 eggs). She will "sit" or "set" on the nest, protesting or pecking in defence if disturbed or removed, and she will rarely leave the nest to eat, drink, or dust-bathe. While brooding, the hen maintains the nest at a constant temperature and humidity, as well as turning the eggs regularly during the first part of the incubation. To stimulate broodiness, an owner may place many artificial eggs in the nest, or to stop it they may place the hen in an elevated cage with an open wire floor.

At the end of the incubation period (about 21 days), the eggs, if fertile, will hatch. Development of the egg starts only when incubation begins, so they all hatch within a day or two of each other, despite perhaps being laid over a period of two weeks or so. Before hatching, the hen can hear the chicks peeping inside the eggs, and will gently cluck to stimulate them to break out of their shells. The chick begins by "pipping"; pecking a breathing hole with its egg tooth towards the blunt end of the egg, usually on the upper side.

It will then rest for some hours, absorbing the remaining egg yolk and withdrawing the blood supply from the membrane beneath the shell (used earlier for breathing through the shell). It then enlarges the hole, gradually turning round as it goes, and eventually severing the blunt end of the shell completely to make a lid. It crawls out of the remaining shell, and its wet down dries out in the warmth of the nest. The hen will usually stay on the nest for about two days after the first egg hatches, and during this time the newly hatched chicks live off the egg yolk they absorb just before hatching. Any eggs not fertilized by a rooster will not hatch, and the hen eventually loses interest in these and leaves the nest.

After hatching, the hen fiercely guards the chicks, and will brood them when necessary to keep them warm, at first often returning to the nest at night. She leads them to food and water; she will call them to edible items, but seldom feeds them directly. She continues to care for them until they are several weeks old, when she will gradually lose interest and eventually start to lay again.

Modern egg-laying breeds rarely go broody, and those that do often stop part-way through the incubation. However, some "utility" (general purpose) breeds, such as the Cochin, Cornish and Silkie, do regularly go broody, and they make excellent mothers, not only for chicken eggs but also for those of other species—even those with much smaller or larger eggs and different incubation periods, such as quail, pheasants, turkeys or geese. Chicken eggs can also be hatched under a broody duck, with varied success.

Poultry Farming

More than 50 billion chickens are reared annually as a source of food, for both their meat and their eggs. Chickens farmed for meat are called broiler chickens, whilst those farmed for eggs are called egg-laying hens. In total, the UK alone consumes over 29 million eggs per day. Some hens can produce over 300 eggs per year. Chickens will naturally live for 6 or more years, but broiler chickens typically take less than six weeks to reach slaughter size. For laying hens, they are slaughtered after about 12 months, when the hens' productivity starts to decline, by which point they are normally infirm and have lost a significant amount of their feathers, and their life expectancy has been reduced from around 7 years to less than 2 years.

The vast majority of poultry are raised using intensive farming techniques. According to the Worldwatch Institute, 74 percent of the world's poultry meat, and 68 percent of eggs are produced this way. One alternative to intensive poultry farming is free range farming.

Friction between these two main methods has led to long term issues of ethical consumerism. Opponents of intensive farming argue that it harms the environment, creates human health risks and is inhumane. Advocates of intensive farming say that their highly efficient systems save land and food resources due to increased productivity, stating that the animals are looked after in state-of-the-art environmentally controlled facilities.

In part due to the conditions on intensive poultry farms and recent recalls of large quantities of eggs, there is a growing movement for small scale micro-flocks or 'backyard chickens'. This involves keeping small numbers of hens (usually no more than a dozen), in suburban or urban residential areas to control bugs, utilise chicken waste as fertilizer in small gardens, and of course for the high-quality eggs and meat that are produced.

Artificial Incubation

Incubation can successfully occur artificially in machines that provide the correct, controlled environment for the developing chick. The average incubation period for chickens is 21 days but may depend on the temperature and humidity in the incubator. Temperature regulation is the most critical factor for a successful hatch. Variations of more than 1 °F (1.8 °C) from the optimum temperature of 99.5 °F (37.5 °C) will reduce hatch rates.

Humidity is also important because the rate at which eggs lose water by evaporation depends on the ambient relative humidity. Evaporation can be assessed by candling, to view the size of the air sac, or by measuring weight loss. Relative humidity should be increased to around 70% in the last three days of incubation to keep the membrane around the hatching chick from drying out after the chick cracks the shell. Lower humidity is usual in the first 18 days to ensure adequate evaporation.

The position of the eggs in the incubator can also influence hatch rates. For best results, eggs should be placed with the pointed ends down and turned regularly (at least three times per day) until one to three days before hatching. If the eggs aren't turned, the embryo inside may stick to the shell and may hatch with physical defects. Adequate ventilation is necessary to provide the embryo with oxygen. Older eggs require increased ventilation.

Many commercial incubators are industrial-sized with shelves holding tens of thousands of eggs at a time, with rotation of the eggs a fully automated process. Home incubators are boxes holding from half a dozen to 75 eggs; they are usually electrically powered, but in the past some were heated with an oil or paraffin lamp.

Chicken Eggs as Food

Chicken eggs are widely used in many types of dishes, both sweet and savory, including many baked goods. Eggs can be scrambled, fried, hard-boiled, soft-boiled, pickled, and poached. The albumen, or egg white, contains protein but little or no fat, and can be used in cooking separately from the yolk. Egg whites may be aerated or whipped to a light, fluffy consistency and are often used in desserts such as meringues and mousse. Ground egg shells are sometimes used as a food additive to deliver calcium. Some people prefer to just have a female, and raise it for the eggs.

Chickens as Food

The meat of the chicken, also called "chicken", is a type of poultry meat. Because of its relatively low cost, chicken is one of the most used meats in the world. Nearly all parts of the bird can be used for food, and the meat can be cooked in many different ways. Popular chicken dishes include roasted chicken, fried chicken, chicken soup, Buffalo wings, tandoori chicken, butter chicken, and chicken rice. Chicken is also a staple of many fast food restaurants.

Chickens as Pets

Chickens are sometimes kept as pets and can be tamed by hand feeding, but roosters can sometimes become aggressive and noisy, although aggression can be curbed with proper handling. Some have advised against keeping them around very young children. Certain breeds, however, such as silkies and many bantam varieties are generally docile and are often recommended as good pets around children with disabilities. Some people find chickens' behaviour entertaining and educational.

Chicken Diseases and Ailments

Chickens are susceptible to several parasites, including lice, mites, ticks, fleas, and intestinal worms, as well as other diseases. Despite the name, they are not affected by chickenpox, which is generally restricted to humans.

Some of the common diseases that affect chickens are shown below:

Name	*Common Name*	*Caused by*
Aspergillosis		fungi
Avian influenza	bird flu	virus
Histomoniasis	Blackhead disease	protozoal parasite
Botulism		toxin
Cage Layer Fatigue		mineral deficiencies, lack of exercise
Campylobacteriosis		tissue injury in the gut
Coccidiosis		parasites
Colds		virus
Crop Bound		improper feeding
Dermanyssus gallinae	Red mite	parasite
Egg bound		oversised egg
Erysipelas		bacteria
Fatty Liver Hemorrhagic Syndrome		high-energy food
Fowl Cholera		bacteria
Fowl pox		virus
Fowl Typhoid		bacteria
Gallid herpesvirus 1 or Infectious Laryngotracheitis		virus
Gapeworm	Syngamus trachea	worms
Infectious Bronchitis		virus
Infectious Bursal Disease	Gumboro	virus

Contd...

Name	*Common Name*	*Caused by*
Infectious Coryza		bacteria
Lymphoid leukosis		Avian leukosis virus
Marek's disease		virus
Moniliasis	Yeast Infection	
or Thrush	fungi	
Mycoplasmas		bacteria-like organisms
Newcastle disease		virus
Necrotic Enteritis		bacteria
Omphalitis	Mushy chick disease	umbilical cord stump
Peritonitis		Infection in abdomen from egg yolk
Prolapse		
Psittacosis		bacteria
Pullorum	Salmonella	bacteria
Scaly leg		parasites
Squamous cell carcinoma		cancer
Tibial dyschondroplasia	speed growing	
Toxoplasmosis		protozoal parasite
Ulcerative Enteritis		bacteria
Ulcerative pododermatitis	Bumblefoot	bacteria

Chickens in Religion and Mythology

In Indonesia the chicken has great significance during the Hindu cremation ceremony.

A chicken is considered a channel for evil spirits which may be present during the ceremony. A chicken is tethered by the leg and kept present at the ceremony for its duration to ensure that any evil spirits present during the ceremony go into the chicken and not the family members present. The chicken is then taken home and returns to its normal life.

In ancient Greece, the chicken was not normally used for sacrifices, perhaps because it was still considered an exotic animal. Because of its valour, the cock is found as an attribute of Ares, Heracles, and Athena. The alleged last words of Socrates as he died from hemlock poisoning, as recounted by Plato, were "Crito, I owe a cock to Asclepius; will you remember to pay the debt?", signifying that death was a cure for the illness of life.

The Greeks believed that even lions were afraid of cocks. Several of Aesop's Fables reference this belief.

In the New Testament, Jesus prophesied the betrayal by Peter: "Jesus answered, 'I tell you, Peter, before the rooster crows today, you will deny three times that you know me.'" (Luke 22:34) Thus it happened (Luke 22:61), and Peter cried bitterly. This made the cock a symbol for both vigilance and betrayal.

Earlier, Jesus compares himself to a mother hen when talking about Jerusalem: "O Jerusalem, Jerusalem, you who kill the prophets and stone those sent to you, how often I have longed to gather your children together, as a hen gathers her chicks under her wings, but you were not willing." (Matthew 23:37; also Luke 13:34).

In many Central European folk tales, the devil is believed to flee at the first crowing of a cock.

In traditional Jewish practice, a kosher animal is swung around the head and then slaughtered on the afternoon before Yom Kippur, the Day of Atonement, in a ritual called kapparos. A chicken or fish is typically used because it is commonly available (and small enough to hold). The sacrifice of the animal is to receive atonement, for the animal symbolically takes on all the person's sins in kapparos. The meat is then donated to the poor. A woman brings a hen for the ceremony, while a man brings a rooster. Although not actually a sacrifice in the biblical sense, the death of the animal reminds the penitent sinner that his or her life is in God's hands.

The Talmud speaks of learning "courtesy towards one's mate" from the rooster (Eruvin 100b). This might refer to the fact that when a rooster finds something good to eat, he calls his hens to eat first.

The chicken is one of the Zodiac symbols of the Chinese calendar. Also in Chinese religion, a cooked chicken as a religious offering is usually limited to ancestor veneration and worship of village deities. Vegetarian deities such as the Buddha are not one of the recipients of such offerings. Under some observations, an offering of chicken is presented with "serious" prayer (while roasted pork is offered during a joyous celebration). In Confucian Chinese Weddings, a chicken can be used as a substitute for one who is seriously ill or not available (e.g. sudden death) to attend the ceremony. A red silk scarf is placed on the chicken's head and a close relative of the absent bride/groom holds the chicken so the ceremony may proceed. However, this practice is rare today.

A cockatrice was supposed to have been born from an egg laid by a rooster, as well as killed by a Rooster's call.

Chickens in History

An early domestication of chickens in Southeast Asia is probable, since the word for domestic chicken (**manuk*) is part of the reconstructed Proto-Austronesian language. Chickens, together with dogs and pigs, were the domestic animals of the Lapita culture, the first Neolithic culture of Oceania.

The first pictures of chickens in Europe are found on Corinthian pottery of the 7th century BC. The poet Cratinus (mid-5th century BC, according to the later Greek author Athenaeus) calls the chicken "the Persian alarm". In Aristophanes's comedy *The Birds* (414 BC) a chicken is called "the Median bird", which points to an introduction from the East. Pictures of chickens are found on Greek red figure and black-figure pottery.

In ancient Greece, chickens were still rare and were a rather prestigious food for symposia. Delos seems to have been a centre of chicken breeding.

The Romans used chickens for oracles, both when flying ("ex avibus", Augury) and when feeding ("auspicium ex tripudiis", Alectryomancy). The hen ("gallina") gave a favourable omen ("auspicium ratum"), when appearing from the left (Cic.,de Div. ii.26), like the crow and the owl. For the oracle "ex tripudiis" according to Cicero (Cic. de Div. ii.34), any bird could be used, but normally only chickens ("pulli") were consulted. The chickens were cared for by the pullarius, who opened their cage and fed them pulses or a special kind of soft cake when an augury was needed. If the chickens stayed in their cage, made noises ("occinerent"), beat their wings or flew away, the omen was bad; if they ate greedily, the omen was good.

In 249 BC, the Roman general Publius Claudius Pulcher had his chickens thrown overboard when they refused to feed before the battle of Drepana, saying "If they won't eat, perhaps they will drink." He promptly lost the battle against the Carthaginians and 93 Roman ships were sunk. Back in Rome, he was tried for impiety and heavily fined. In 161 BC, a law was passed in Rome that forbade the consumption of fattened chickens. It was renewed a number of times, but does not seem to have been successful. Fattening chickens with bread soaked in milk was thought to give especially delicious results. The Roman gourmet Apicius offers 17 recipes for chicken, mainly boiled chicken with a sauce. All parts of the animal are used: the recipes include the stomach, liver, testicles and even the pygostyle (the fatty "tail" of the chicken where the tail feathers attach).

The Roman author Columella gives advice on chicken breeding in his eighth book of his treatise on agriculture. He identifies Tanagrian, Rhodic, Chalkidic and Median (commonly misidentified as Melian) breeds, which have an impressive appearance, a quarrelsome nature and were used for cockfighting by the Greeks. For farming, native (Roman) chickens are to be preferred, or a cross between native hens and Greek cocks. Dwarf chickens are nice to watch because of their size but have no other advantages.

Per Columella, the ideal flock consists of 200 birds, which can be supervised by one person if someone is watching for stray animals. White chickens should be avoided as they are not very fertile and are easily caught by eagles or goshawks. One cock should be kept for five hens. In the case of Rhodian and Median cocks that are very heavy and therefore not much inclined to sex, only three hens are kept per cock. The hens of heavy fowls are not much inclined to brood; therefore their eggs are best hatched by normal hens. A hen can hatch no more than 15-23 eggs, depending on the time of year, and supervise no more than 30 hatchlings. Eggs that are long and pointed give more male, rounded eggs mainly female hatchlings.

Per Columella, chicken coops should face southeast and lie adjacent to the kitchen, as smoke is beneficial for the animals. Coops should consist of three rooms and possess a hearth. Dry dust or ash should be provided for dust-baths.

According to Columella, chicken should be fed on barley groats, small chick-peas, millet and wheat bran, if they are cheap. Wheat itself should be avoided as it is harmful to the birds. Boiled ryegrass (*Lollium* sp.) and the leaves and seeds of alfalfa (*Medicago sativa* L.) can be used as well. Grape marc can be used, but only when the hens stop laying eggs, that is, about the middle of November; otherwise eggs are small and few. When feeding grape marc, it should be supplemented with some bran. Hens start to lay eggs after the winter solstice, in warm places around the first of January, in colder areas in the middle of February. Parboiled barley increases their fertility; this should be mixed with alfalfa leaves and seeds, or vetches or millet if alfalfa is not at hand. Free-ranging chickens should receive two cups of barley daily. Columella advises farmers to slaughter hens that are older than three years, because they no longer produce sufficient eggs.

Capons were produced by burning out their spurs with a hot iron. The wound was treated with potter's chalk.

For the use of poultry and eggs in the kitchens of ancient Rome see Roman eating and drinking.

Chickens were spread by Polynesian seafarers and reached Easter Island in the 12th century AD, where they were the only domestic animal, with the possible exception of the Polynesian Rat (*Rattus exulans*). They were housed in extremely solid chicken coops built from stone.

Chickens in South America

An unusual variety of chicken that has its origins in South America is the araucana, bred in southern Chile by Mapuche people. Araucanas, some of which are tailless and some of which have tufts of feathers around their ears, lay blue-green eggs. It has long been suggested that they predate the arrival of European chickens brought by the Spanish and are evidence of pre-Columbian trans-Pacific contacts between Asian or Pacific Oceanic peoples, particularly the Polynesians and South America.

In 2007, an international team of researchers reported the results of analysis of chicken bones found on the Arauco Peninsula in south central Chile. Radiocarbon dating suggested that the chickens were Pre-Columbian, and DNA analysis showed that they were related to pre-historic populations of chickens in Polynesia. These results appeared to confirm that the chickens came from Polynesia and that there were transpacific contacts between Polynesia and South America before Columbus's arrival in the Americas.

However, a later report looking at the same specimens concluded:

> *A published, apparently pre-Columbian, Chilean specimen and six pre-European Polynesian specimens also cluster with the same European/Indian subcontinental/Southeast Asian sequences, providing no support for a Polynesian introduction of chickens to South America. In contrast, sequences from two archaeological sites on Easter Island group with an uncommon haplogroup from Indonesia, Japan, and China and may represent a genetic signature of an early Polynesian dispersal. Modelling of the potential marine carbon contribution to the Chilean archaeological specimen casts further doubt on claims for pre-*

Columbian chickens, and definitive proof will require further analyses of ancient DNA sequences and radiocarbon and stable isotope data from archaeological excavations within both Chile and Polynesia.

Drinking and Feeding Systems

Feeding Systems

There are several different systems available for delivery and distribution of feed to broilers. Since feed constitutes the major share of total production cost, wastage should be an important consideration in the choice of system.

There are three major systems available:

- Automatic pan feeders: 1 pan per 65 birds; 33 cm pan diameter.
- Chain feeders: 2.5 cm per bird; 80 birds per metre of track.
- Round, hanging tube feeders: 65 birds per tube; 38 cm diameter base.

Automatic pan feeding systems have become the industry standard due to advantages of low feed wastage, ease of height adjustment, preservation of pellet quality, and reliability. As a number of different pan feeder designs are available, feeder heights should be set according to manufacturers' recommendations. Distance between the feeder lines should be not more than 2.5 metres. This ensures that all birds have adequate access to feed. Level of feed within the feeder should be adjusted to a height that minimises wastage. If possible, the feed supply system should be allowed to empty at least once a day. This eliminates the presence of stale food and therefore reduces the risk of contamination and the growth of micro-organisms.

Drinking Systems

It is essential that fresh water is available to the broiler flock at all times and that it is free of contamination. The drinking systems chosen must be capable of delivering the water efficiently to all birds with the minimum of spillage. To ensure that the flock is receiving sufficient water, each day, the ratio of water to feed consumed should be monitored. When the ratio of water volume (ml or l) to feed weight (g or kg) remains close to 1.8:1 (1.6:1 for nipple drinkers), only then can it be assumed that the birds are consuming sufficient water. Birds will drink more water at high ambient temperatures. Water requirement increases by approximately 6.5% per degree as temperature exceeds 21°C. Water consumption will vary with feed consumption.

Nipple Drinkers

Nipple systems provide water with lower levels of bacterial contamination than conventional open systems. They have become the standard in modern broiler production.

General recommendations for the management of nipple systems are:

- 12 birds per nipple. This should be reduced to 9-10 per nipple for birds weighing 2.75 kg or more.
- Nipple height should be monitored daily and adjusted as appropriate. At day old, nipples should be placed at chick eye level. From day 2 onward, while drinking, the back of the chick should form an angle of 45° with the floor.
- Litter, under and around the drinker lines, should be level to allow all birds to have equal access to water.
- Drinker lines should be level to a avoid spillage.
- Individual nipples should be checked regularly to confirm that access is available to birds through 360° (i.e. from all directions). Faulty nipples will reduce birds' access to drinking water. Nipples should be activated and checked by hand before placement to ensure all nipples are working.
- Water pressure should be set according to manufacturers' specifications.
- Nipple lines should be flushed and sanitised weekly.

Bell Drinkers

- When whole house brooding is practiced, a minimum of 6 bell drinkers should be provided per 1000 chicks.
- Drinkers should be distributed evenly throughout the house so that no broiler is more than 2 m from water.
- As a guide to level, water should be 0.6 cm below the top of the drinker until 7-10 days and there should be 0.6 cm of water in the base of the drinker from 10 days onwards.
- The height at which the bell drinkers are suspended should be checked and adjusted daily, so that the lip of the bell is level with the broilers' backs from 7 days onwards.

Agritech

Comparative evaluation of feed conservation in fibreglass and metal silos during summer and winter time

Agritech srl has entrusted the Faculty of Agriculture of the Università Cattolica del Sacro Cuore, in cooperation with Cerzoo, the Research Center for Zootechnics and Environment of Piacenza (Italy) with a comparative study on the performance of some own-manufactured fibreglass silos and other silos in galvanised metal, produced by Chore Time. The study covers a period of time going from July 21st 2008 to February 2nd 2009.

Aim of the Study

The aim of the study was the comparative analysis of the conservation of bulk feed in mealy form stored in fibreglass and galvanised metal silos during summer and winter time.

Compared Samples

1. *Storage silos:* The study was carried out on a total of 6 silos of 6 ton capacity each, supplied by Agritech, 3 of which in fibreglass (VTR) and 3 in metal (MET). The silos were installed in pairs (VTR-MET) in the same environmental conditions with regard to the exposure to the sun.
2. *Stored feed:* The feed selected for the test was commercial compound feed in mealy form, taken from the same production batch, and stored in the silos in the same quantity. The silos were only loaded by 2/3 of their real capacity in order to reproduce the normal conditions of use in a standard farm, where silos are progressively emptied. To put in evidence eventual effects connected with non-ideal storage and environmental conditions typical of summer, some vegetal oil was added to the main feed as lipidic integration.
3. *Period:* The study was carried out from July 2008 to February 2009.
4. Surveys: During the 4-month summer test following surveys were carried out:
   a) *Temperature (T°):* Using Min. and Max. thermometers, the values of T° MAX., T° MIN. and T° INSTANT in the external environment, the air temperature inside the silos (the empty volume between the cover of the silos and the surface of the feed) and the temperature of the feed (with a thermometer being placed in the first 10 cm of the bulk) were measured on alternate days.
   b) The amount of peroxides released over 20 days, based on feed samples taken from the superior and inferior part of

each silo. During the 3-month winter test following surveys were carried out:

c) *Temperature (T°):* Using Min. and Max. thermometers, the values of T° MAX., T° MIN. and T° INSTANT in the external environment, the air temperature inside the silos (the empty volume between the cover of the silos and the surface of the feed) and the temperature of the feed (with a thermometer being placed within the first 10 cm of the bulk) were measured on alternate days.

5. *Results:* As far as the summer period is concerned, it was noticed that:

   a) There are remarkable differences between the temperature of the air inside the silos and that of the stored feed in relation to the building material of the silos. These differences, that arise from the majority of the surveys, are statistically significant, and they are indicated in the charts by some asterisks corresponding to the date of the survey. As one can observe, the large amount of significant marks proves that fibreglass silos can stand thermal fluctuations and control both the air and the feed temperature better than metal silos.

   In fibreglass silos (VTR), instant, minimum and maximum temperatures resulted to be better than in metal silos (MET) in the majority of the surveys.

   Particularly, the Authors observed higher MAX. T° of the feed in the upper part of the bulk and of the air inside metal silos, as reported in chart No. 3 Delta T° between VTR and MET.

   As reported, these differences reach up to over 8°C in the air inside silos, with registered MAX. T° over 45°C in the metal silos. The temperature of feed in the upper part of silos has reached temperatures between 35°C and 45°C in the metal silos, while in the fibreglass silos the highest temperature never exceeded 35°C (this peak was only reached in three surveys). The temperature deltas (Ä) comparatively registered in the temperatures of the feed stored in VTR or MET silos confirm the better performance of fibreglass silos, with a difference in favour of fibreglass up to 7°C.

b) Regarding the amount of released peroxides registered in the upper and lower part of the feed bulk, the fibreglass silos show a lower rate of peroxides and consequently a lower oxidation of the lipids contained in feed than metal silos (...).

Regarding the winter period (November 2008 – February 2009), the Authors had following results:

a) "Max. temperature of the air inside the silos: all temperature data registered in fibreglass silos are statistically inferior to those registered in metal silos. The average difference measured during the 15 test-weeks is about - 106,32%. These values in fibreglass silos tend to match with those of the environmental temperature in the "hottest" weeks, while, similarly to what happens with the minimum temperatures, the lowest ("coldest") values tend to be inferior to those measured in the outside environment".

b) "Max. temperature of the feed inside silos: in the majority of the surveys, the MAX. T° of the feed stored inside fibreglass silos show values that are much inferior to those registered in metal silos.

"Finally, the results obtained in the second part of the test, which integrate and complete the results of the summer test, prove that also in the winter months the temperature registered in the air and feed inside silos is the parameter which is subject to the most significant variations, and that it largely depends on the building material of the silos. In fibreglass silos, both the temperature of the inside air and the temperature of the stored feed bulk are averagely inferior to the temperatures registered in metal silos".

This e-mail address is being protected from spambots. You need JavaScript enabled to view it

Big Dutchman

ReproMatic and FluxxBreeder – new feeding system especially for broiler breeders. ReproMatic is a feeding system developed by Big Dutchman to ideally meet the particular requirements of broiler breeder management. Only this system allows all birds to receive feed immediately and simultaneously. It combines the advantages of chain and pan feeding. The rugged feed chain is used as conveying system.

The feed channel with chain allows for a high filling level and consequently a very high conveying capacity. FluxxBreeder is Big Dutchman's newly developed feed pan that can be used in rearing for day-old to death production and in broiler breeder production.

The main aim in rearing is for all hens of a flock to reach laying maturity at the same time. A uniform flock can only develop, if all birds have sufficient space to feed. Moreover, pans must be filled simultaneously at the same speed and to the same level to allow all birds of a flock to receive the same amount of feed during restricted feeding.

Features of FluxxBreeder in the rearing phase:

ReproMatic-pullets: The same view with pullets. 16 "true" feeding spaces, which means 60 % more birds per running metre of feeding system as compared to a linear trough;

- 360° flooding mechanism that ensures a high feed level in the pan, especially in the first days of rearing;
- spin-n-lock system allows for simple, one-handed adjustment of the feed level;
- flat pan for an ideal start of day old chicks, a good distribution of feed and reduced feed losses;
- the elevated feed channel and rotatable pan provide the birds with enough freedom of movement;
- eight wings on the outer cylinder of the pan prevent feed losses as lateral feed spillage is not possible;
- the integrated volume reducer allows for small feed rations, thus ensuring fast and simultaneous filling of all pans;
- after the birds have been moved out, the pans can easily be cleaned with a high-pressure cleaner, for drying the pan bottom can simply be opened;
- excess cleaning water or disinfectants can easily drain off through additional holes in the pan bottom;
- the conveying system consists of a feed channel equipped with the Challenger feed chain, thus allowing the transport of large amounts of feed with a high conveying capacity (2 t/h);
- smooth feed saving lip prevents bruises and feed losses;
- sliding shut-off to close individual pans;
- ideal illumination of the pan due to openings in the pan top.

Amacs is Big Dutchman's sophisticated and modern management system for the control of the entire feed supply but also for climate, water supply, bird weighing and lighting. Amacs allows to control a low-maintenance batch. Since the feed lines can be filled when they are suspended there is no need for a large hopper. Amacs has a modular design and can be used for small and large houses alike, as it can be adapted to the individual situation on a farm. Amacs allows for ongoing data collection, real-time control and monitoring of individual barns or entire farm complexes – all this from virtually any location in the world.

Chore-Time

Brock Europe offers poultry drinking and feeding systems.

Chore-Time Europe B.V. offers drinker and feeder options for commercial poultry producers.

Chore-Time's popular RELIA-FLOW® Nipple Drinker offers producers a reliable flow rate consistent with the way birds actually drink. This university-tested drinking system features robust, precision-machined, stainless steel parts in the flow-control area of the valve which resist wear and retain their shape for long life and consistent, reliable flow. The system includes a one-piece valve for easy field replacement in retrofit applications. Chore-Time also offers a basic drinker model with its STEADI-FLOW® Nipple Drinker. For turkey producers, Chore-Time offers the ADVANTI-FLOW® Poult Drinker. The ADVANTI-FLOW Drinker features pockets in the drinker's disc that hold attractive beads of water. The pockets and the disc's scalloped edge work together to control the direction of water flow. The disc also provides leverage for easier triggering of the nipple drinker by young birds. Dual catch cups with rounded edges for bird comfort maximise water consumption while helping keep floors dry.

Chore-Time recently introduced a new drinking system regulator which is easier for users to install, manage and maintain. Designed for long, reliable service, the regulator requires less pressure to seal than other regulators, resulting in less wear on the sealing mechanism. Simple-to-use, top-mounted knobs activate the "regulate" or "flush" modes for each drinker line, while a bottom knob is used to adjust the operating water column at each regulator.

Also available from Chore-Time is a new, large-diameter, folding stand tube which folds to protect both the tube and the ceiling during house clean out. The rigid PVC stand tube requires no spring for support, making the water column easier to see and helping the tube to stay cleaner. The large-volume pipe removes more air from the watering system than smaller diameter tubes.

Chore-Time offers poultry producers convenient control of the water pressure levels in all nipple drinker lines in the house with its updated PDS™ Controls. With Chore-Time's PDS (Pneumatic Drinking System) Controls, users can change the pressure in all lines in the house or flush all lines from one remote location.

One of Chore-Time's most popular feeding systems is its Model C2® PLUS Pan Feeding System. The feeder is available in a standard model, as well as models with a shallow pan to help with starting birds or a model with extended fins to enhance feed flow for harder-flowing feed types. The chick-friendly 14-spoke grill design includes feed-saving features to help maximise feed conversion throughout the growing cycle.

Chore-Time's also offers its unique REVOLUTION® Poultry Feeder. Unlike conventional feeder pans with flood windows, the REVOLUTION Feeder includes a Rotary Feed Gate which provides full control of both flood and final feed levels without needing to depend on pan height or gravity to operate. On the floor or in the air, the Rotary Gate gives the REVOLUTION Feeder complete control of both flood and final feed levels.

Chore-Time Europe B.V. is an affiliate of CTB, Inc. Based in Milford, Indiana (U.S.A.), CTB, Inc. is a leading global designer, manufacturer and marketer of systems and solutions for the poultry, pig, egg production, and grain industries. Its products and services are "Helping to Feed a Hungry World®" through improved efficiency and air quality management in the care of poultry and livestock as well as in grain storage, handling, conditioning and drying.

Founded in 1952, CTB has been dedicated to "Leadership Through Innovation®" throughout its history. The company operates from multiple locations in various countries around the world and serves its customers through a worldwide network of independent dealers and distributors.

8

Egg Production

Types of Poultry Enterprises

Backyard poultry production is at the subsistence level of farming. Birds live free range and hatch their own eggs. Their diet is supplemented with crop waste or food leftovers. The labour involved in backyard poultry production is part-time. Farm flock production is slightly more specialized. Eggs are hatched at a separate location where the hatch and the sexing of the birds are controlled. Commercial poultry farm production involves full-time labour and is geared toward producing on a sufficient scale for the sale of both eggs and poultry meat. Specialized egg production consists of separating poultry for meat and egg production. In the egg producing plant, specialized employees oversee specific aspects of egg production. Integrated egg production is the most advanced enterprise and involves full mechanization and automation of the egg production cycle including battery egg laying, temperature controls, scientific feeding and mechanized egg collection methods.

Oil Barrel-Charcoal

All of the above poultry-keeping methods are used in the developing world, but the majority of the enterprises are backyard poultry and farm flock production. The poultry and egg sectors are highly fragmented. Most of the production is carried out by a large number of farmers, each with a very small flock. The greater part of produce is sold in markets close to the farms. Day-old chicks are usually obtained from local hatcheries licensed by international hybrid breeding companies. Farmers or cooperatives of farmers may choose between varieties of chickens for egg production and meat production.

The small chicks can be either naturally or artificially brooded. If artificially brooded, small chicks must be placed in a separate house from laying chickens and it is necessary to protect the chicks from predators, diseases and catching colds. This stage of brooding lasts for eight weeks. In the first four weeks of life, small chicks need to be housed in a brooding box.

Kerosene Brooder

Storm Lantern Brooder

After the first month, small chicks are removed from the brooder box and placed in the brooder house. At two months of age, the chicks enter the grower stage which lasts until they are five months (20 weeks) old. Growers may either be housed separately from small chicks or continue to be reared in brooder-cum-grower houses. It is important to properly manage the growers as their reproductive organs develop during this period and this will affect their egg production capacity in the future. When the growers reach 18 weeks of age they are moved to laying houses and begin to lay eggs, which are, however, small and unmarketable. It is not until they are 21 weeks old that the growers reach their commercial laying stage. Layers may be placed in intensive, semi-intensive or free-range types of housing. The choice of housing is determined by climate, type of production desired and the farmer's financial resources. Some examples of laying houses are shown on the next two pages.

Factors Affecting Egg Production

Typically, a layer's production cycle lasts just over a year (52-56 weeks). During the production cycle many factors influence egg production; therefore, the cycle must be managed effectively and efficiently in order to provide maximum output and profitability. The following factors influence egg production.

Breed: The breed of the laying bird influences egg production. Management and feeding practices, however, are the key determining features for egg production.

Mortality Rate: Mortality rate may rise due to disease, predation or high temperature. The mortality rate of small chicks (up to eight weeks of age) is about 4 percent; that of growers (between eight and 20 weeks of age) is about 15 percent; and that of layers (between 20 and 72 weeks of age) is about 12 percent. The average mortality rate of a flock is from 20 to 25 percent per year.

Types of Laying Houses

Age: Birds typically begin producing eggs in their twentieth or twenty-first week and continue for slightly over a year. This is the best laying period and eggs tend to increase in size until the end of the egg production cycle.

Body Weight: In general, optimum body weight during the laying period should be around 1.5 kg, although this varies according to breed. Underweight as well as overweight birds lay eggs at a lower rate. Proper management and the correct amount of feed are necessary in order to achieve optimum body weight.

Laying House: The laying house should be built according to local climatic conditions and the farmer's finances. A good house protects laying birds from theft, predation, direct sunlight, rain, excessive wind, heat and cold, as well as sudden changes in temperature and excessive dust. If the climate is hot and humid, for example, the use of an open house construction will enable ventilation. The inside of the house should be arranged so that it requires minimum labour and time to care for the birds.

Lighting Schedule: Egg production is stimulated by daylight; therefore, as the days grow longer production increases. In open houses, found commonly in the tropics, artificial lighting may be used to increase the laying period. When darkness falls artificial lighting can be introduced for two to three hours, which may increase egg production by 20 to 30 percent.

In closed houses, where layers are not exposed to natural light, the length of the artificial day should be increased either in one step, or in a number of steps until the artificial day reaches 16 to 17 hours, which will ensure constant and maximized egg production. Effective day length should never decrease during the laying period.

Feed: Free-range hens will produce more meat and eggs with supplemental feed, but only if they are improved breeds or crossbreeds. The selection of local hens is done on the basis of resistance and other criteria rather than feed utilisation for production. Fresh and clean water should always be provided, as a layer can consume up to one-quarter of a litre a day.

Culling: Culling is the removal of undesirable (sick and/or unproductive) birds, from the flock. There are two methods of culling:

- mass culling, when the entire flock is removed and replaced at the end of the laying cycle; and

- selective culling, when the farmer removes individual unproductive or sick birds.

Culling enables a high level of egg production to be maintained, prevents feed waste on unproductive birds and may avert the spreading of diseases.

Climate: The optimal laying temperature is between 11° and 26° C. A humidity level above 75 percent will cause a reduction in egg laying.

Management Factors: Effective and efficient management techniques are necessary to increase the productivity of the birds and consequently increase income. This entails not only proper housing and feeding, but also careful rearing and good treatment of the birds.

Temperature and its Effects on Egg Production

Temperature (°C)	*Effects*
11-26	Good production.
26-28	Some reduction in feed intake.
28-32	Feed consumption reduced and water intake increased; eggs of reduced size and thin shell.
32-35	Slight panting.
25-40	Heat prostration sets in, measures to cool the house must be taken.
40 and above	Mortality due to heat stress.

When the temperature rises above 28° C the production and quality of eggs decrease. Seasonal temperature increases can reduce egg production by about 10 percent.

Vaccination and Disease Control: Diseases and parasites can cause losses in egg production.

Some of the diseases are as follows:

- bacterial: tuberculosis, fowl typhoid
- viral: Newcastle, fowl plague
- fungal: aspergillosis
- protozoan: coccidiosis
- nutritional: rickets, perosis.

Some of the parasites are:

- external: lice, mites
- internal: roundworms, tapeworms.

Vaccinations are administered to birds by injection, water intake, eye drops and spraying. Clean and hygienic living quarters and surroundings may eliminate up to 90 percent of all disease occurrences.

Collection of Eggs

Frequent egg collection will prevent hens from brooding eggs or trying to eat them and will also prevent the eggs from becoming damaged or dirty.

Egg Production Cycle

Birds usually start to lay at around five months (20-21 weeks) of age and continue to lay for 12 months (52 weeks) on average, laying fewer eggs as they near the moulting period. The typical production cycle lasts about 17 months (72 weeks) and involves three distinct phases, as follows.

Phase 1: Small chicks or brooders— This phase lasts from 0 to 2 months (0-8 weeks) during which time small chicks are kept in facilities (brooder houses) separate from laying birds.

Phase 2: Growers— This phase lasts about 3 months, from the ninth to the twentieth week of age. Growers may be either housed separately from small chicks or continue to be reared in brooder-cum-grower houses. It is important to provide appropriate care to the growers particularly between their seventeenth and twentieth week of age as their reproductive organs develop during this period.

Phase 3: Layers— Growers are transferred from the grower house to the layer house when they are 18 weeks old to prepare for the laying cycle. Birds typically lay for a twelve-month period starting when they are about 21 weeks old and lasting until they are about 72 weeks old.

Production Schedule in Temperate Climate (100 birds)

Age of flock (in weeks)	*% of flock laying*	*No. of birds laying*	*No. of eggs produced per week*
21	5	5	20
22	10	10	40
23	18	18	72
24	34	34	136
25	52	52	208
26	65	65	260

Contd...

Age of flock (in weeks)	*% of flock laying*	*No. of birds laying*	*No. of eggs produced per week*
27	74	74	296
28	84	84	336
29	88	88	352
30	92	92	368
31	94	94	376
32-39	88	88	352
40-47	83	83	332
48-59	77	77	308
60-64	73	73	292
65-70	70	70	280

Clearly, egg production requires planning for costs as well as for profit generation and for meeting market demand. Planning involves not only the number of eggs laid by the flock over a period of time, but also when to hatch chicks to replace birds with diminishing laying capacity.

Production Planning

On average a bird produces one egg per day. Furthermore, not all birds start to lay exactly when they are 21 weeks old. Planning is therefore required for egg production to be constant so as to meet market demand. In areas where the climate is hot and humid, commercial hybrid laying birds produce on average between 180 and 200 eggs per year. In more temperate climates birds can produce on average between 250 and 300 eggs per year.

As shown in the third column, for 100 birds at 21 weeks of age only five would actually be laying. In the fourth column the actual number of eggs produced is shown. On average a bird produces 208 eggs over a twelve-month period, which is a weekly production rate of four eggs per bird. At 21 weeks of age 20 eggs are produced (five birds produce four eggs each) and at 22 weeks 40 eggs are produced, etc. Egg production rises rapidly and then starts to fall after 31 weeks of age. When less than 65 percent of the flock are laying eggs (71 weeks of age), it may become uneconomical to retain birds. Feed costs and sales of culled birds for meat must be considered as well as prices for eggs. In some instances when egg prices are high it may be viable to delay culling birds until only 45 percent of the flock is still laying eggs (78 weeks of age).

As indicated on the chart, the first layer flock was hatched at 0 weeks to become productive after 21 weeks. The second flock of layers was hatched at the 21st week to be ready to lay after the 41st week, as the first layer flock starts to diminish production. This type of production entails having flocks of birds of different age groups. Clean and hygienic living quarters and surroundings are essential to control disease. There should be no more than three or four different flock age groups present at one time. The mortality rate on average is between 20 and 25 percent. This means that if one wants 100 birds to lay, it may be necessary to buy between 120 and 125 small chicks.

Production Costs and Profits

Records should be kept of costs incurred during the operation and of proceeds from the sale of eggs. Costs must be covered by the sales of eggs. The difference between the proceeds from the sales and costs incurred represents profit.

Brooder-grower Stage

The costs to be considered are not only those concerned with the birds during the laying period, but also those incurred in the brooder and grower stage during which time no eggs are being produced. The brooder-cum-grower stage lasts about five months (0-20 weeks).

Laying Birds

Once the costs for the brooder-cum-grower stage have been calculated, it will be possible to calculate costs for the laying birds. Calculations may be made on a daily, weekly or monthly basis. However, the most useful calculations are made at the end of the laying cycle. Daily, weekly or monthly calculations give approximate indications of costs and relative profits or losses. The main concern for farmers during this period is probably whether or not the proceeds from the sale of eggs cover feed and rearing costs.

Feed cost is generally estimated to be about 75 percent of the production cost of eggs. Comparing feed and rearing costs and egg proceeds for a week or a month may give an indication of profitability or loss. A farmer would have to subtract the cost of feed for a week from the proceeds for the total number of eggs sold that week. Furthermore, the rearing costs (expenses incurred before the birds start laying) should be amortized. This can be calculated by dividing the total rearing costs by the laying period. If rearing costs are US$ 10 and the laying period is 52 weeks, cost per week for rearing is US$ 0.19.

Costs and Income for the Laying Cycle

Calculations for the laying cycle (52 weeks) are more accurate and enable the farmer to determine whether the egg laying enterprise is running at a profit or a loss.

Expenses for Rearing

Costs	*US$*
Chicks (total number of chicks multiplied by price per chick)	
Feed (total kg of feed multiplied by price per kg)	
Housing	
Equipment	
Labour	
Vaccinations	
Mortality	
Loan	
Various	
Total costs	

Weekly costs and sales

	US$
a) Eggs sold	
b) Feed used	
c) Rearing costs	
a minus b and c =	

Costs and Income for a Production Cycle*

Costs	US$
Rearing	
Houses	
Equipment	
Feed	
Labour	
Vaccinations	
Mortality	

Contd...

Costs	**US$**
Various expenses	
Total costs	
Income	
Sale of eggs	
Sale of culled birds	
(Sale of manure)	
Total income	
Profit	

* This table does not include marketing costs.

Costs: When calculating costs for the laying cycle, the main expenditures to consider are:

- rearing-rearing brooders until they become layers;
- housing-building or maintaining laying house and brooder house;
- equipment-the cost of miscellaneous items such as feeders, buckets, etc.;
- feed-total feed used during the year;
- labour-labour costs incurred to manage birds;
- vaccinations-medicines and veterinary visits;
- mortality-loss of laying birds due to disease, etc.; and
- various expenses-lighting, water, etc.

Income: When calculating income for the laying cycle, the earnings to consider derive from:

- the sale of eggs;
- the sale of culled birds after the first cycle of production; and
- where applicable, manure sold as fertilizer.

Initially, capital is required to start an enterprise; proceeds from the sales of eggs should, however, provide funds to continue with the business before the end of the first laying cycle. Indeed, three months after point of lay (30-31 weeks of age), when the birds should normally have reached peak production, the proceeds from the sale of eggs should be sufficient to operate the business on a revolving fund basis. The three-month period is sufficiently long even for the low producing birds or those that peak late.

Guidelines for Improved Household Poultry Production

Constraints: The main limitation to improved household poultry production is the extremely high loss of birds before they reach maturity caused by inadequate nutrition and disease. This loss means that a high proportion of all the eggs laid have to be kept for replacement stock leaving little, if any, surplus for sale or consumption. The main causes of loss are:

- Poor nutrition is the major cause of loss and predisposes birds to disease, poor immune response to vaccines and predation.
- Disease, especially the highly infectious viral Newcastle Disease (ND), which is believed to be endemic in most rural flocks. Clinically the disease is cyclic and occurs at times of climatic and nutritional stress. The virulent (velogenic) strain common in Africa and Asia can, but not always, cause up to 80 percent mortality in unvaccinated chickens. Fowl cholera (pasteurellosis), coccidiosis, Gumboro disease (infectious Bursal disease) and fowl pox can also, to a lesser extent, cause problems in rural flocks.

Poor, or non-existent housing, is also a major cause of high losses. Without being able to confine birds at night, it is almost impossible to catch and vaccinate them, although new types of ND vaccine can be administered in the feed. Shelter can also provide protection for young birds against predators and can ensure that all the eggs are laid in the proper place and not lost.

The majority of indigenous breeds or strains of chicken/fowl have evolved to survive under harsh conditions where they largely have to fend for themselves. Such hardiness, however, is at the expense of higher levels of productivity and they are less able to exploit the advantages of improved management, nutrition, etc., than breeds with a greater genetic potential for egg production and feed conversion (growth).

Potential: Improved management and disease control can have a substantial impact on household economies.

Under traditional management the majority of eggs are hatched to ensure sufficient replacements with only the male birds being sold or consumed. Reduced losses will ensure that more birds could be successfully reared and, assuming the extra birds can be properly fed, this will allow more eggs to be collected and consumed or sold as a regular source of income.

Potential Interventions

The basis for any improvement programme will be improved husbandry, notably housing, nutrition and disease control, primarily Newcastle Disease. Subsequent interventions would concentrate on further improving nutrition and the introduction of improved breeds/strains.

Improved Feeding

Most household flocks rely on scavenging and household scraps and, depending on conditions, this is usually adequate for survival and a low level of production. However, inadequate nutrition, exacerbated by marked seasonal fluctuations, is a major predisposing factor to disease and high mortality. As investments are made in improved animal health, housing and, especially if improved birds are to be introduced, then attention must be given to diet supplementation or feeding a complete diet in the case of totally confined birds.

Conventional feed materials such as maize, wheat, barley, oilcakes, fishmeal, etc., are rarely available to the back-yard producer. In many developing countries these are in short supply and even compounded feeds may be of dubious quality. For household production systems, however, there are usually a wide range of locally available feedstuffs that can be used in addition to household scraps. These include: surplus/broken or second grade grains (cereals, maize, sorghum and millet); roots and tubers (sweet potatoes, cassava, etc.), green material (legumes and leaf meals, sweet potato vines, etc.), residues and agro-industrial by-products (bran, rice polishings, oilseed cakes, etc.). Unless a complete balanced ration is available, the ability to free range is important to allow the birds to feed on insects and worms, green material, etc., so that they can balance their essential amino acids, mineral, vitamin, as well as energy requirements. Where appropriate, improved feeding systems (troughs, etc.) should be supplied to reduce wastage. Access to clean water is always essential and a source of calcium (ideally ground oyster shell) is highly recommended.

Control of Newcastle Disease (ND) and other Health Constraints

Effective vaccines have been available against most strains of ND for a long time. However, there are a number of issues that need to be addressed:

1. Until recently, the potency of vaccines was highly sensitive to temperature which meant that the provision of an effective vaccine at village level required a 'cold chain' of refrigerators,

cool boxes, etc., from the manufacturing laboratory through to the farm. The majority of vaccines are still highly sensitive to temperature and fall within this class.

2. Conventional vaccines are sold in large dose vials, usually 1 000 doses, aimed at the commercial producer but unsuitable for use at the village or household level.
3. Village flocks are usually small, scattered and multi-aged which makes them difficult to target by mass vaccination campaigns. Catching free range, often semi-feral chickens to vaccinate them individually has always proved difficult.
4. Vaccination of a multi-age flock has to be undertaken on a continuous basis (monthly) to be effective.

A new 'heat stable' oral vaccine has been developed and widely tested in Asia and Africa. The primary advantage is that it no longer requires a complete cold chain to maintain its potency. Queensland University in Australia has made available free to laboratories in developing countries a seed virus, designated I2, to those who wish to explore the possibilities of vaccine production. This opens the door for producing with intermediate levels of technology, the fresh (not freeze-dried) vaccine at regional laboratories for use within a few weeks of production.

In addition, a commercial V4 vaccine is also available, but not in large quantities and it remains expensive. Potentially these vaccines offer the possibility of overcoming the problems of transport, storage and the difficulty of catching individual chickens.

They are not, however, available everywhere, and applying the vaccine to feeds is not without problems. The question of who produces the vaccine remains an issue and experience has shown that projects may be able to introduce the technology but often production ceases once external inputs are removed.

Conventional vaccines remain a viable option if there is a reliable 'cold chain', if housing is provided that allows the birds to be caught easily and if sufficient numbers of owners participate, making the use of large vials economic. There is often little difference, however, in cost between 200-and 1 000-dose vials. A major problem with the larger vials is to find and catch 1 000 village chickens within the two hours or so that these 'old' heat sensitive vaccines remain viable. Almost all birds in rural flocks are infected with a variety of internal parasites which cause reduced growth rate, weight loss and lower egg production. Strategically timed treatment(s) with inexpensive anthelmintics (e.g. fenbendazole and other benzimidazoles) given in the feed can easily eliminate the majority of these parasites.

Improved Housing

The basic aim should be to provide simple (using local materials wherever possible) yet secure housing for the birds at night. Approximately 0.1m2 (1ft2) should be adequate per bird. Housing should provide: perches for birds to roost on; access to clean water; a creep feed for chicks; and, nest boxes for laying and brooding. Location should be close to the house to deter theft and preferably raised off the ground to provide protection from predators and to reduce dampness. The shelter should have easy access to allow for catching the birds with the minimum of disturbance. Such housing can usually be provided cheaply using local materials (timber, mud, thatch, etc.); however, more complex designs may require more expensive sawn timber and wire netting.

Improved Breeds

Once standard levels of husbandry (housing, feeding and disease control) have been achieved, improving the genetic potential of the birds offers the next step in increasing productivity. One strategy is to use local birds to incubate and rear higher egg-producing breeds.

Two Choices are Available: The introduction of pure-bred, dual-purpose breeds (e.g. the Rhode Island Red or Australorp) or the commercial hybrids, which are usually selected either for meat (broiler) or egg production. Traditionally, the dual-purpose breeds have been the exotic breeds of choice, the exception has been the White Leghorn, a laying breed that has proved unsatisfactory in adapting to village conditions. Obtaining grandparent stock of these breeds is becoming increasingly difficult and expensive. Some commercial companies now offer a more hardy, dual-purpose type of hybrid bird that could be used in certain situations.

Securing a regular source of healthy birds from well managed hatcheries can be problematic. Traditionally, government services have maintained poultry farms with imported parent stock and have supplied day-old-chicks (DOCs) or point-of-lay (POLs) birds to farmers. However, as with so many state run operations, there are real problems in managing such enterprises efficiently. Lack of working capital and staff incentives have resulted in most of them operating at a very low level of productivity and at a financial loss. The alternative of placing such activities in the private sector should be encouraged. Initially this may involve a phased approach through increasing cost-recovery to full privatization of government services. Non-governmental organizations can have a role in providing skills, start-up loans, etc., to assist private entrepreneurs in establishing themselves. Wherever possible the incubation, brooding, rearing and production of hatching eggs can be undertaken by separate specialized producers within the village.

In many developing countries improved birds have to be imported. There are a number of options that can be considered:

- *Importing grandparent stock to produce parent stock in the country.* This requires high levels of management, a regular supply of quality inputs, and a sufficient demand for parent stock.
- *Importing parent stock as either fertile eggs or day-old chicks to supply commercial birds for distribution.* This is usually the most economic option if acceptable levels of production can be maintained.
- *Importing commercial fertile eggs or day-old chicks for direct supply to farmers.* This option might be feasible in establishing a programme but it is costly. Although the full costs involved in producing DOCs locally from parent stock may exceed the cost of importing commercial DOCs if management and performance is low. With full cost recovery, these costs will have implications for the financial viability of the enterprise that must be understood.

There are other issues that also need to be considered. The indiscriminate distribution of imported breeds could have long-term adverse effects in diluting the advantageous traits in the indigenous breeds, especially broodiness in local hens.

There is potential for improving locally adapted breeds by selection. Virtually all the indigenous breeds have not been subjected to any selection process, other than natural selection. The consequence is that there is a large variation in production traits (i.e. number of eggs laid, etc.) between individuals in the overall population. By identifying and selecting the top performers for a given trait, and given the chicken's short generation interval, it would be possible to make substantial gains in genetic potential within the existing production environment. However, care must be taken since some traits are genetically negatively correlated i.e. broodiness and egg production. The logistical constraints in successfully implementing such a programme are formidable.

Institution support. The promotion and development of producer groups as the basis for self-sufficiency should be supported through training (technical and business management) and start-up capital in the form of goods or services. Involvement and support for the private sector in the provision of goods and services should be encouraged and, initially, this would involve the introduction of cost recovery for government goods and services that provide a 'private' rather than a 'public' benefit.

Marketing Quality Eggs

Quality Criteria

Quality determines the acceptability of a product to potential customers. The quality of eggs and their stability during storage are largely determined by their physical structure and chemical composition. It is important therefore that those concerned with the handling of eggs are knowledgeable about this information in order to understand why eggs need to be treated in specific ways and to have a rational basis for day-to-day marketing decisions.

Composition and Attributes of Eggs

An egg consists of shell, membrane, albumen or white and yolk.

The shell: The shell of an egg has a rigid yet porous structure. The porous shell has great resistance to the entry of micro-organisms when kept dry and considerable resistance to the loss of moisture by evaporation. The colour of the shell, which may be white or brown depending on the breed of the laying chicken, does not affect quality, flavour, cooking characteristics, nutritional value or shell thickness.

Shell Membrane: Inside the shell there are two membranes. The outer membrane is attached to the shell, the inner membrane is attached to the albumen or egg white. These two membranes provide a protective barrier against bacterial penetration.

Air Space: An air space or air cell is a pocket of air usually found at the large end of the egg interior between the outer membrane and the inner membrane. This air cell is created by the contraction of the inner contents while the egg cools and by the evaporation of moisture after the egg has been laid. The air cell increases in size as time passes.

Egg Albumen or White: The albumen of the egg is composed of the outer thin albumen and the inner firm or thick albumen. The outer thin albumen spreads around the inner firm albumen. The inner firm albumen in high quality eggs stands higher and spreads less than the outer thin albumen.

White Fibrous Strips: These are twisted, cord-like strands of egg white, known as chalazae, which hold the yolk in position. Prominent thick chalazae indicate high quality and freshness.

Yolk: The yolk is almost spherical and is surrounded by a colourless membrane. The colour of the yolk varies with the type of feed given to the laying hen.

If the laying hen is fed on maize, for example, the yolk will become a bright yellow. The colour of the yolk does not affect the nutritional content.

Egg Weight: The weight of eggs varies widely depending on many factors such as the breed, the age of the layer and environmental temperature. In Africa, for example, the egg weight may range from 35 to 65 grams, while in Europe it may range from 45 to 70 grams. As a layer gets older the weight of the eggs increase as can be seen in the following figure.

The components of an egg weighing 60 grams are made up as follows:

- yolk (29%)-17.4 g
- white (61.5%)-36.9 g
- shell (9.5%)-5.6 g.

Nutritional Value

Eggs are a good source of high quality protein. They provide important sources of iron, vitamins and phosphorus. As a nutritional source of vitamin D, eggs rank second only to fish liver oils. Eggs are low in calcium, which is discarded in the shell, and contain very little vitamin C.

Eggs provide a unique and well-balanced source of nutrients for persons of all ages. Hard-cooked egg yolks are of great nutritional value as a major supplementary source of iron for infants. When children reach one year of age they may also be given egg whites. Eggs contain significant nutritional value, which is essential during rapid body growth, and therefore are excellent food for young children and teenagers.

Low caloric value, ease of digestibility and high nutrient content make eggs valuable in many therapeutic diets for adults. During convalescence, when bland diets may be required, eggs provide a good nutritious diet. For older people, whose caloric needs are lower, eggs are an easy, inexpensive and nutritious food to prepare and eat. Availability, modest cost, ease of preparation, popular taste appeal and low caloric value give eggs a primary advantage for human nutritional needs.

Shell Quality: Texture, Colour, Shape and Condition

The ideal shape of an egg as established by tradition and by practical considerations. Shell quality characteristics that must be considered are as follows:

- cleanliness
- soundness (unbroken)
- smoothness
- shape.

The two most desirable shell qualities, cleanliness and soundness, are largely controlled by the production and handling of eggs. Eggs with shell defects should be removed from eggs destined to the retail trade. Clearly, consumers have adverse reactions to cracked or dirty eggs. Even if the cracks in an egg are only visible when candling, the micro-cracks may have serious consequences on quality. These eggs may be sold locally and possibly only a few hours after lay.

When the membrane is broken as well as the shell, the contents of the eggs can leak, and therefore the only practicable market outlet is sale as egg pulp. If the eggs are dirty, for example, with blood or faeces, consumers will react unfavourably to them.

Although shell colour is no indication of quality, consumers in some markets may prefer white eggs or brown eggs. In such circumstances, it is advisable to sort eggs by shell colour.

Yolk and Albumen Quality

In quality eggs the yolk should be round, firm and stand up well, and be yellow in colour. There is often prejudice against very pale or deeply coloured yolks, however, there are some exceptions. In some Italian markets, for example, red yolks are a strong selling point. The yolk should have a pleasant, mild egg odour and flavour and should be surrounded by a large amount of upstanding thick white with only a small amount of thin white. The egg white should have the normal slightly green-yellow colour, though it may be slightly cloudy in appearance.

Consumers are generally very critical of any abnormal conditions in the egg yolk and white. Factors that may cause loss of quality are as follows:

- natural factors
- temperature
- humidity
- time
- handling
- storage
- tainting.

Natural factors, for example, can be blood spots, which may range from small specks to a square centimetre in size. They may vary in colour from light grey to bright red and may be found in the yolk or in the egg white. "Blood eggs", with blood diffused throughout the white or spread around the yolk, are not commonly found and are generally rejected by the consumer.

Deterioration

The changes that occur in eggs stored for a week to ten days at a temperature between 27° and 29° C are comparable to those that occur in similar eggs in cold storage for several months at a temperature of-1° C. In advanced stages of deterioration, the thick white may disappear entirely and the yolk may enlarge to the point where its membranes are so weakened that it breaks when the egg is opened. Changes in odour and flavour take three or four weeks at a temperature of 21° C, or six to seven weeks at a temperature of 10° C to become noticeable to the ordinary consumer.

Temperature, humidity, air movement and storage time can all have adverse effects on interior quality. These factors, if not controlled, can cause loss of moisture in eggs. Loss of water through the porous shell will mean loss of weight. A loss of weight of two to three percent is common in marketing eggs and is hardly noticeable to consumers. However, enlarged air cells and a decreased size of egg contents become noticeable when losses exceed this extent. Coating eggs with oil and other substances and storing them at low temperatures and high humidity may control moisture loss. The best conditions for storage are at a temperature of about-1° C and relative humidity between 80 and 85 percent. At a temperature of 10° C, lower relative humidity is needed, between 75 and 80 percent. At all temperatures there is the risk of mould spoilage where the relative humidity is too high. Packaging materials that are too dry or are excessively moist and absorbent will also accentuate evaporation losses.

The contents of eggs when just laid are usually sterile and contain few organisms capable of causing spoilage even when the shells are slightly dirty or stained. The main cause of spoilage by bacteria is the washing of dirty eggs before marketing. When the egg is washed, organisms from water-usually bacteria-can penetrate the shell. Once inside they multiply and eventually spoil the egg, causing green, black and red rots. Even when eggs become wet without any cleaning process, for example, by condensation after removal from refrigerated storage into a warm temperature, conditions may be favourable for the penetration of micro-organisms and rotting may follow. When eggs are kept dry, no such way is provided for bacteria to penetrate the shell. Mould spores normally present on eggshells may, if sufficient time elapses, germinate and grow, penetrating the shell and causing spoilage. Generally this occurs only when eggs are in cold storage for several months or more under conditions of high humidity (above 85 percent). It can occur, however, at any temperature if the humidity is sufficiently high and the holding time long enough.

Eggs can easily be tainted by strong odours from kerosene, gasoline, diesel oil, paint and varnish, and by such fruit and vegetables as apples, onions and potatoes. Special care must therefore be taken in storage, packaging materials and transport facilities used.

Quality Maintenance

Maintaining fresh egg quality from producer to consumer is one of the major problems facing those engaged in marketing eggs. Proper attention to production, distribution and point-of-sale phases are of vital importance in maintaining egg quality.

Production Factors

The main production factors that affect quality maintenance are the following:

- breed
- age
- feed

- management
- disease control
- handling/collecting eggs
- housing.

Breed: The breed of the laying hen affects shell colour; for example, Leghorns produce white eggs, while Rhode Island Reds produce brown eggs. The following egg quality factors are partly inherited: shell texture and thickness, the incidence of blood spots and the upstanding quality and relative amount of thick albumen. Though it may not always be possible, a consistent policy of selection for breeds by egg producers can bring noticeable improvements to quality.

Age: Birds typically begin producing eggs in their twentieth or twenty-first week and continue for slightly over a year. This is the best laying period and eggs tend to increase in size until the end of the egg production cycle. Birds lay fewer eggs as they near the moulting period. In the second year of lay, eggs tend to be of lower quality.

Feed: Egg quality and composition derive primarily from what a layer is fed. In terms of taste, for example, eggs laid by hens fed on fishmeal will have a "fishy" taste. The type of feed will also influence the shell of an egg and the colour of the yolk. Layers must be kept away from certain plant foods if egg colour defects are to be avoided. These may include cottonseed meal and the foliage of the *sterculiaceae* and *malvaceae* such as mallow weed. Regular access to fresh or high-quality dehydrated green feed helps birds to produce eggs with a uniform yellow yolk. Yellow maize, alfalfa meal, and fresh grass provide good pigment sources for a normal yellowish-orange yolk colour.

Management: Good general management of the laying flock can improve egg quality. If birds are treated correctly and not put under conditions of stress they will produce properly.

Disease Control: Diseases have an effect on egg quality. Infectious bronchitis and Newcastle disease, for example, will cause birds to lay eggs with poor quality shells and with extremely poor quality albumen. Many of the birds continue to lay poor quality eggs even after recovery. Effective vaccines should be administered.

Handling/Collecting Eggs: Frequent collection is essential each day in order to limit the number of dirty and damaged eggs and also to prevent the hens from eating the eggs. Careful handling is necessary in order to avoid breakage.

Laying House: The number of dirty eggs produced can be reduced significantly by providing good housing and clean nests for the layers. Cleaning and hygiene operations should be carried out frequently.

Measures to Prevent Deterioration During Marketing

Temperature: By far the most effective method of minimizing deterioration of quality in eggs is to keep them at temperatures below 13° C. Eggs should never be left standing in the sun or in a room that gets very hot at some point in the day, but should be moved into shaded, well-ventilated rooms and underground cellars as quickly as possible. Various methods to prevent deterioration by temperature are shown below.

1. A simple method is covering eggs with green leaves, so as to reduce temperature.
2. A method commonly used is that of putting eggs in a porous pot where the outside of the pot is kept damp. Great care should be taken, however, to avoid the excess use of water, which could trickle down to the bottom of the pot damaging the eggs at the bottom.
3. Eggs can be kept in a wide-mouthed earthen pot that is buried in the ground up to half of its height. The inside of the pot is lined with a thin layer of grass to prevent the eggs being spoiled by excess moisture. Eggs are placed in the pot as soon as they are collected and the top covered with a thin cloth to facilitate the exchange of air. A layer of sand and earth is spread around the earthen pot and water is sprinkled on it frequently during the day. The eggs are turned once a day to prevent the internal yolk of the egg from sticking to one side of the eggshell. Such a system may reduce the egg temperature by 8° C below the temperature outside the pot.
4. Another method that can be used which is ideal for dry climates makes use of the cooling effect of evaporation. Baskets of eggs are stored in a small wooden or wire-frame cupboard. A water tray is kept on top of the box and pieces of sacking are placed in the tray and arranged so that they hang on all sides of the box. More elaborate versions with arrangements for a steady dripping of water on to the sacking can be developed. In humid areas such devices would be less useful. The maintenance of egg quality in wet tropical areas is extremely difficult without refrigeration.

5. Refrigerated storerooms can be used if electricity is available. An example of a refrigerated storeroom is shown below. If refrigerated storehouses are not economically viable, the use of electric fans may be appropriate.

Refrigerated Storeroom

The cooler should be shaded from direct sunlight. It is most efficient when air circulates *freely* round it and does not give good results in a closed room. The cooling effect can be intensified by blowing air through damp sacking with an electric fan.

Producers, wholesalers and retailers should move eggs to consumers as quickly as possible to minimize the risk of spoilage. The importance of avoiding delays at all stages in the distribution channel cannot be overemphasized and should be the primary consideration determining marketing arrangements.

Treatment of Dirty Eggs: Some eggs will inevitably have dirty shells. For the purpose of appearance, washing is the most effective and simplest method of removing dirt and stains from the shell surface. The water, however, may contain bacteria that could penetrate the porous eggshell causing it to decay. Odourless detergent-sanitizing substances should be used in the water to wash eggs, but these may be difficult to obtain. Eggs can be submerged in clean hot water (water temperature should be around 38° C); however, this may cause thermal cracks in the eggshell and internal expansion of the egg content. It is better to avoid washing eggs altogether. Using dry abrasives for scraping and brushing may be the optimal solution. In using this method, care should be taken to avoid removing excessive shell material, which will weaken the shell and increase the rate of evaporation.

The cooler should be shaded from direct sunlight. It is most efficient when air circulates *freely* round it and does not give good results in a closed room. The cooling effect can be intensified by blowing air through damp sacking with an electric fan.

Shell Oiling: Coating eggs with a thin film of oil greatly reduces losses by evaporation, especially where eggs are in cold storage for several months or are held at temperatures above 21° C. Special odourless, colourless, low-viscosity mineral oils should be used. Where eggs must withstand high temperatures, they should be oiled from four to six hours after lay. If eggs are to be stored at a temperature of 0° C, they should be oiled 18 to 24 hours after lay. Eggs can be oiled by hand dipping wire baskets or by machine. The temperature of the oil should be at least 11° C above that of the eggs. Before the oil is reused it should be heated to a temperature of 116° C to prevent bacteria survival and then be filtered. The oil reservoirs should be cleaned properly. In terms of appearance oiled eggs differ from other eggs only in the slight shine left on the eggshells by the more viscous oils.

Inducements for Quality Maintenance

Provision of effective incentives for the adoption of quality maintenance procedures is the function of the marketing system. It must provide some means whereby egg quality can be appraised and a system of purchasing premiums and deductions applied accordingly. Methods of assessing quality of eggs are discussed below.

Grading and Standardization

Grading and standardization consist of arranging produce into a number of uniform categories according to physical and quality characteristics of economic importance. It is a process of identification, classification and separation. The advantages of grading and standardization are as follows:

- Different grade eggs may be sold to different customers. Customers willing to pay more for high quality eggs will be served. On the other hand, eggs with micro-cracks or small blood spots may be sold to bakeries.
- Setting and maintaining a reliable standard creates consumer confidence in the product and a favourable reputation. This will enable buyers (wholesalers, retailers, exporters, consumers) to purchase a reliable product that they recognize and may well avoid inspection and disputes.
- The ability to furnish an accurate description of eggs in storage may help in obtaining credit.

Grade Specifications

The value factors most generally appreciated in eggs are internal quality, appearance and soundness of shell, size and colour.

Most egg marketing systems find it advantageous to adopt grading practices that:

- eliminate inedible and defective eggs;
- separate eggs into high and lower acceptable categories; and
- establish uniform weight classifications.

The various egg sizes according to weight used in the United States are as follows:

- Jumbo = 70 g and above
- Extra large = 65-70 g
- Large = 56-65 g
- Medium = 49-56 g
- Small = 42-49 g
- Peewee = 35-42 g.

Of course different size specifications and quality factors vary from country to country; for example, the various sizes according to weight used in Africa are as follows:

- Large = 65 g and above
- Medium = 55-65 g
- Small = 45-55 g.

Quality Specifications Development

A simple set of quality specifications might be set up as follows.

First Grade: The shell must be clean, unbroken and practically normal in shape and texture. The air cell must not exceed 9.5 mm in depth and may move freely, but not be broken and bubbly. The yolk may appear off-centre, but only slightly enlarged, and may show only slight embryonic development. No foreign objects may be present.

Second Grade: The shell must be unbroken, but may be somewhat abnormal in shape and texture. Only slight stains and marks are permitted. The yolk may appear dark and enlarged and may show embryonic development, but not at the blood vessel stage and beyond. Blood spots less then 6mm in diameter are permitted.

Third Grade: Other edible eggs are permitted, that is, those that are not rotted, mouldy or musty. Also, those eggs that are not incubated to blood vessel stage, and those not containing insects, worms or blood spots 6mm in diameter, are permitted.

Interior Quality

The most accurate test of interior quality is the break-out method-open the egg on to a flat glass surface and compare the appearance of the yolk. In marketing, however, the above method can only be used on a sample basis. A method that does not require egg breaking is more appropriate.

Candling

Candling is the only method of testing eggs for quality, internally and externally, without breaking them. It consists of inspecting an egg with a beam of light that makes the interior quality visible. A very simple form of candling is placing a candle in a dark room and positioning an egg in front of the flame and looking at the interior quality. If electricity is available, a light bulb can be placed in the box, otherwise a paraffin lamp or candle can be enclosed in a casing. The hole in the box should be about 3 cm in diameter, sufficient for egg sizes ranging from 40 to 70 grams. The light will shine out of the hole making the interior egg quality visible.

Candling Method: Pick four eggs to be candled and put two in each hand. Place the first egg near the candling box hole with the large end of the egg held against the light, and with the axis at a 45° angle so that the egg has light shining through it.

Twirl the egg so as to observe defects which otherwise might not be observed. If first egg candled is free of defects, roll the first egg back in to the palm of the hand. Meanwhile, the second egg in the other hand should be brought to the light and examined. While the second egg is up against the light, the third egg in the other hand should be brought into candling position. Although the beginner will soon learn to detect such things as cracked shells and bloodspots, considerable training is needed before internal quality can be estimated with reasonable accuracy. The main interior quality points to be observed in candling can be summarized as follows:

Yolk: The judgement of internal quality is based mainly on the visibility, ease of movement and shape of yolk. Common yolk faults are the following:

- *Sided*-displaced to an appreciable extent from its normal central position.
- *Stuck*-on twirling the egg, it may be found that the yolk is stuck to the inner shell membrane.
- *Patchy*-uneven in colour, including defects sometimes described as "heat spots."
- *Abnormal in shape*-flattened or irregular and in extreme cases may be broken and dispersed in the white.
- *Discoloured*-of a dark or greyish appearance often with a very distinct outline.

- *Embryonic development*-first shows as a dark halo round the germ cell near the centre of the yolk and later as thin blood vessels and a bright blood ring.

White: In practice, the quality of egg white is judged by the degree of movement of the yolk and by the definition of its outline. Common faults in egg white are as follows:

- *Discoloured*-definitely tinted grey, yellow, green or brown.
- *Cloudy*-muddy or streaky. Usually this condition indicates potential rot, but washing an egg in very hot water can cause a similar appearance.

Air Cell: The depth of the air cell is a rough indication of the age of the egg and there is often a relation between this depth and the internal quality.

Hence, the depth of the air cell is taken into account in candling, but other indications of quality are given equal weight. The air cell may be:

- *Large*-exceeding 6 mm in depth.
- *Running*-if the air cell is broken, one or more air bubbles will be found in the white. If the air cell has forced its way between the two shell membranes, bubbles will move around the shell when the egg is candled. A running air cell, however, may be caused by rough handling and should not exclude the egg from a high-grade class.
- *Ringed*-the air cell is very large, sharply defined and with grey or brown edges.

Other common defects of internal quality that may be found are listed here:

- *Blood spots*-clots or streaks of blood in the white or adhering to the yolk.
- *Blood egg*-blood is diffused throughout the white or spread around the yolk.
- *Meat spots*-fatty material, fleshy or liver-like that may be found floating freely in the white, embedded in the chalazae or attached to the yolk.
- *Staleness*-in most cases the air cell is abnormally large, clearly defined and often ringed. As a rule the yolk is sided and its outline clearly defined.

- *Mould growth*-usually grey or black in colour, but can occasionally be pinkish, found on the outside and inside of the shell or shell membranes.
- *Rot*-usually violet, green, red or blue in colour. The early stages of a rot are less easy to detect, but any egg with a streaky, turbid white should be rejected. The egg may have an unpleasant smell even if unbroken.
- *Taint*-the egg has an abnormal odour.

Shell condition: Weak, rough, mouldy, cracked and deformed shells may be detected as eggs are picked up for candling. But with candling small or micro-cracks on the eggshell can be seen. Another method of verifying shell soundness is that of gently hitting two eggs together (belling). A dull sound instead of clear clinking indicates a cracked egg.

Developing a Standardized System

Any system used to grade the quality and weight of eggs is only effective if it ensures that consumers obtain eggs of the quality and kind they want. This requires that three conditions be met.

1. The initial classification must be correct.
2. There should be no appreciable deterioration between time of grading and time of sale.
3. The consumers should have a clear guide to the quality of produce they are purchasing.

In developing a grading system, the following preparations are vital.

- Study thoroughly the pattern of production, consumption and trade.
- Work out grade specifications in close consultation with those traders who would be likely to take advantage of them.
- Qualified inspectors are needed to ensure that conformity with the grades indicated is upheld. Producers and packers who accept their inspection can be authorized to apply approved grade stamps.
- Legislation should be enacted to prevent the possibility of misleading labelling. Certain standards may be made obligatory, for example, minimum standards to protect consumers from unwholesome and dangerous foods.

- Finally, if consumers are to take full advantage of a grading system, the grade indications should be clear and easily understood.

It is important that eggs are not allowed to deteriorate below the grade indicated before they reach the consumer. Under conditions of high humidity with temperatures of 32° C and above, eggs being distributed may undergo considerable deterioration in only a few days. Under such circumstances, the inspection of eggs held for more than three or four days must be clearly and responsibly assigned.

Equipment and Candling Layout

In order to grade and pack eggs with consistent accuracy, speed and economy, it is essential that adequate facilities be provided. These include a semi-darkened room, without stray illumination, that is well equipped with bench and shelf space and with good facilities for handling of eggs and recording results. The candling device, if possible, should be adjustable and mounted at such a height that the beam of light when emerging horizontally from the device arrives at about the height of the operator's elbow.

The lamp or light bulb in the candling box should be kept clean. An operator should use the same candling apparatus every day to help minimize errors. Each operator must have enough room to move freely and handle packing materials and boxes. Efficiency can be increased by installing partitions between candling benches, arranging supplies of packing materials conveniently and by providing for the easy disposal of damaged materials.

Floors and walls should have smooth, hard non-reflecting surfaces and be rounded at intersections. Thorough cleanliness is essential to prevent bad odours. Good ventilation should be provided. The candling bench should be designed to particular needs. Where there are only a few candling benches, operators themselves can obtain eggs and supplies from nearby stocks and carry away completed egg cases. As the volume handled increases, however, it usually becomes more efficient to assign a special worker to service the candlers.

Egg Packaging, Transport and Storage

Packaging of Shell Eggs

Nature has given the egg a natural package-the shell. Despite its relative strength, the egg is an extremely fragile product and even with the best handling methods, serious losses can result from shell damage. Economical marketing generally requires that eggs be protected by the adoption of specialized packaging and handling procedures.

Functions of Packaging

Packaging is an important component in delivering quality eggs to buyers. It embraces both the art and science of preparing products for storage, transport and eventually sale. Packaging protects the eggs from:

- micro-organisms, such as bacteria;
- natural predators;
- loss of moisture;
- tainting;
- temperatures that cause deterioration; and
- possible crushing while being handled, stored or transported.

Eggs also need to breathe, hence the packaging material used must allow for the entrance of oxygen. The material used must be clean and odourless so as to prevent possible contamination and tainting. Authentic egg packaging materials can be reused, but careful attention must be paid to possible damage, odours and cleanliness. The packaging must be made to withstand handling, storage and transport methods of the most diverse kind and to protect the eggs against temperatures that cause deterioration and humidity. Finally, consumers like to see what they are buying, especially if it concerns fresh produce.

An egg package should be designed so that the customers not only recognize the product as such, but can also see the eggs they are buying. Many factors must be taken into consideration for packaging eggs. It is important to obtain information regarding the necessary requirements for a particular market, such as:

- quality maintenance;
- storage facilities;
- type of transport;
- distance to be travelled;
- climatic conditions;
- time involved; and
- costs.

Egg Packages

There are many different types of egg packages, which vary both in design and packaging material used.

Type 1: Packing eggs with clean and odourless rice husks, wheat chaff or chopped straw in a firm walled basket or crate greatly decreases the risk of shell damage. The basket has no cushioning material such as straw and therefore damage to the eggs may occur more easily. This kind of packaging may be fit for short distance transport.

Type 2: A very common form of packaging is the filler tray. The fillers are then placed in boxes or cases.

Filler trays are made of wood pulp moulded to accommodate the eggs. They are constructed so that they can be stacked one on top of the other and can also be placed in boxes ready for transport. Filler trays also offer a convenient method for counting the eggs in each box, without having to count every single egg. Usually the standard egg tray carries 36 eggs. Therefore, if a box holds five trays, for example, the box has a total of 180 eggs (36 x 5 = 180). The cases used may be made of sawn wood; however, they are more commonly made of cardboard. When using cardboard cases, special care must be taken in stacking so that excessive weight is not placed on a case at the bottom of a stack. The advantages of using plastic egg fillers are that they can be reused and are washable. The fillers can be covered with plastic coverings and be used as packages for final sale to the buyer. More importantly, however, plastic transparent fillers allow for the inspection of eggs without handling or touching the eggs.

Type 3: Eggs can also be packed in packages that are smaller and specific for retail sale. Each package can hold from two to twelve eggs. It is also possible to pack eggs in small paperboard cases and cover them with plastic film. Egg cases have also been developed from polystyrene. The advantages of using polystyrene are superior cushioning and protection against odours and moisture. The package is also resistant to fungus and mould growth.The use of small cases is restricted by availability and cost considerations. However, small cases are good for retailers and customers. They are easy for the retailers to handle and customers are able to inspect the eggs.

Labelling

Labels are a source of important information for the wholesaler, retailer and consumer and not just pieces of paper stuck onto cartons or boxes. The important facts on the label contain information for buyers concerning the eggs, their size and weight and quality/grade description-AA, A or B. Labels may also indicate the producer, when the eggs were laid, how to store them and their expiration date. Persuading the buyer to purchase the product without tasting, smelling or touching is another function of labelling.

Labels can be either printed directly on cartons or attached to the cartons. The cost of labelling must be taken into consideration.

Costs of Packaging

When calculating the costs of packaging, expenses must be considered for:

- packaging materials;
- labelling;
- labour;
- additional working capital required;
- changing existing facilities (if applicable); and
- packaging machinery (if applicable).

Storage of Eggs

The storage of shell eggs during the main laying season, in order to conserve them for consumption when they are scarce, has been practised for many centuries. For the successful storage of eggs, the following conditions must be met.

- The eggs placed in storage must be clean; they must not be washed or wet.
- Packaging material used should be new, clean and odourless.
- Loss of water due to evaporation should be reduced to a minimum.
- The storage room must be free from tainting products and materials and should be cleaned regularly with odourless detergent sanitizers.
- The storage room must be kept at a constant temperature and humidity must be checked.
- There should be air circulation in the storage room.
- Eggs should be stored so that they are allowed to breathe.
- As far as possible, interior quality should be monitored; there should be a good proportion of thick white, the yolk should stand up well, and the flavour of white and yolk should be good.

If all of the above requirements are to be met, refrigerated storage is necessary.

Cold Storage of Eggs

In the tropics, eggs can deteriorate very quickly unless they are stored at low temperatures. The ideal temperature for storage in such climates is 13°C or lower (usually between 10° and 13° C). Here refrigeration is a necessity for successful commercial storage; however, it may be unavailable or the costs too high.

The most important factors in successful cold storage are as follows:

- The selection and packaging of eggs.
- The equipment and preparation of the cold store.
- Proper temperature, humidity and air circulation.
- Periodic testing for quality.
- The gradual adjustment of eggs to higher temperatures when removed from storage.

The Selection and Packaging of Eggs for Storage: Eggs for storage must be clean, of good interior quality and have a sound shell. If they are to be stored for more then a month, they should be equivalent to the U.S. grade A. Therefore, it is best to candle all eggs before storage. It may also be advisable to take a sample and to break out these eggs as a further quality check. The period of time between laying and storage should not be more than a few days. The eggs should be kept cool during that time.

Packaging materials used for storage should be new, clean, odourless and free from damage. When packaging material is reused, it is extremely important that it is clean, odourless and free from damage. It is important that the material used allow the eggs to "breathe" and to be free from tainting odours. It should also be sturdy in the event that the cases have to be stockpiled on top of one another.

The Equipment and Preparation of the Cold Store: The storage room should have a concrete floor that is washable. Walls and ceilings must also be washable. Wooden buildings have been found to be satisfactory, provided they do not impart foreign odours or flavours to the eggs. The room should be scrubbed thoroughly with hot water and soap or an odourless detergent sanitizer before being used. A final rinse with a hypochlorite solution will help greatly in deodorizing the storeroom. A liberal application of freshly slaked lime to unpainted plaster surfaces will also help. The storage room should be aired and dried out thoroughly after cleaning, then closed up and the refrigeration turned on. It is best to allow several days for the temperature and humidity to stabilize before introducing the eggs.

Proper Temperature, Humidity and Air Circulation: Careful and accurate control of the air condition is essential. A temperature between-1.5° and-0° C is recommended. At a temperature of-2.5° C eggs freeze. The room should be well constructed and insulated and the refrigeration should be capable of maintaining an adequate uniform temperature in all areas. The cases of eggs should be separated by wood-strips and kept well away from the walls so as not to obstruct air circulation. Aisles left for the convenience of handling specific egg cases also help air circulation. Periodic ventilation of the storage room is advisable to promote air exchange.

The relative humidity should be between 80 and 85 percent at a cold storage temperature of-1° C. At cold storage temperatures of about 10° C the relative humidity should be between 75 and 80 percent. In such instances, on average, egg weight loss should not exceed 0.5 percent per month. During the early stages of storage when the packaging material is absorbing moisture at a high rate, the floors should be sprinkled with clean water several times a day. If forced-air circulation is feasible, a controlled temperature water-spray air washer may be used. If the humidity becomes excessive, part of the air can be cycled through a unit containing calcium chloride. Where eggs have been oiled less attention can be paid to the humidity level.

Periodic Testing for Quality: Periodic quality checks are essential if the risk of heavy egg losses is to be avoided. Every month or so a sample of eggs should be selected from the various lots and tested. Usually a sample of about 1 percent of all eggs in storage may be sufficient. For example, if 3 000 eggs are kept in storage, 30 eggs sampled from various egg cases will enable a good estimation of the general quality level of the eggs. If there is evidence of excessive deterioration, it is best to dispose of the eggs quickly, after eliminating those that are unfit for consumption.

The Gradual Adjustment of Eggs to Higher Temperatures: Care must be taken in removing eggs from storage to avoid the condensation of moisture on shells. This is minimized by raising the temperature slowly or by moving the eggs through rooms with intermediate temperatures. If condensation occurs, the eggs should be held under conditions that allow the moisture to evaporate within a day or so. As indicated earlier, eggs should not be stored with products that may taint them. For the long term, eggs are best stored alone, while for the short term they may be kept with dairy products such as milk and mild-flavoured cheese. The average storage life for eggs is between six and seven months.

Economics of Cold Storage

There is a tendency to underestimate the difficulties involved in providing good cold storage facilities and to recommend their installation without adequate investigation of their cost and potential economic return. The following factors should be considered when contemplating cold storage.

- Refrigeration is a complex and highly technical business.
- Capital investment and operating costs must be estimated.
- Potential available business must be appraised.

The potential available business must be appraised as well as its distribution over the different seasons of the year and the costs involved. Egg storage to even out the availability of supplies is likely to provide business for only a part of the year.

The average-not maximum-price difference between the plentiful and scarce seasons must be calculated. If projected returns do not significantly exceed the costs envisaged for storage, there is little incentive for egg traders to make use of storage.

It must be considered that some egg producers, according to their circumstances and possibilities, maintain yearly production through special breeding and feeding programmes and by providing illumination in the hen laying houses. This may even out the rate of egg production throughout the year and hence long-term storage should not be considered.

Layout of Packaging and Storing Facilities for Shell Eggs

A model layout of a packaging and storing room for shell eggs is seen below:

1. Eggs enter the packing/storage facility
2. Temporary store room
3. Candling room
4. Weighing/cleaning room
5. Packaging area

6. Long-term storage
7. Eggs ready for transport.

The layout for the packaging and storing facility is of great importance for efficient and effective management. The various rooms should be kept clean, well ventilated and, where necessary, refrigeration provided. All personnel working in the facility should wear clean outer garments, use caps or head bands and wash their hands when handling eggs and equipment. All equipment used should be clean.

Eggs Enter the Packing/Storage Room: Eggs from production are brought into the packing/storage facility. Eggs can be brought in by hand or by conveyor belt. In intensive egg production, the birds lay eggs that roll out of the cage onto conveyor belts, which transport the eggs directly to the packing/storage facilities.

Temporary Storage Room: Here eggs are stored temporarily before they are moved to the candling room.

Candling Room: Eggs are brought into the candling room, where candlers verify the interior and external quality of eggs. The small squares represent candling benches.

Cleaning/Weighing Room: After candling, eggs are transported by conveyor belts to the cleaning/weighing room. Here eggs are cleaned with abrasives, where possible, and sorted by weight. Usually the size indicates which category eggs should fall into-small, medium or large. This can be done by hand; however, automated weighing machinery is available.

Packaging Area: After weighing, the eggs are taken to the packaging area. Packaging can be done either by hand or automatically by machinery. Each machine is set to pack only predetermined egg weights. For example, machine No.1 packs only 60-gram eggs. If the eggs are below that weight, they will be conveyed to machine No. 2. The eggs are then packed automatically.

After packing, eggs may either be kept in long-term storage or may be ready for immediate transport.

Transport of Eggs

For the successful transport of shell eggs three essential requirements must be met.

1. The containers and packaging materials must be such that the eggs are well protected against mechanical damage.

2. Care should be taken at all stages of handling and transport. Workers handling eggs should be instructed so that they appreciate the need for careful handling. The provision of convenient loading platforms at packing stations, loading depots and railing stations, and handling aids, such as hand trucks and lifts, are of great help.
3. The eggs must be protected at all times against exposure to temperatures that cause deterioration in quality as well as contamination, especially tainting.

The permissible range of temperatures during loading and transport depends on the local climatic conditions and the duration of the journey. Care is needed to avoid excessive shaking, especially where roads are bad. Egg containers should be stacked tightly and tied down securely to minimize movement.

Covers should be used to protect them from the heat of the sun, rain and extreme cold where applicable. Where bicycles are used, a device such as a special carrier suspended on springs may be helpful.

Recommended Temperatures for Loading and Transport

	Transport over 2 or 3 days	*Transport over 5 or 6 days*
Maximum on loading	+6° C	+3° C
Recommended for transport	-1° to + 3° C	-1° to + 1° C
Acceptable for transport	1° to + 6° C	1° to + 3° C

A basic prerequisite for all long-distance transport is that arrangements be made for proper reception, handling and storage at the end of the journey. This is especially important where large lots are delivered to a relatively small market. Without access to suitable storage facilities, the eggs may have to be marketed quickly under adverse climatic conditions, which may cause substantial quality deterioration and price losses. Delivery of high quality eggs over long distances, especially in hot climates, generally calls for refrigeration. Requirements for the successful operation of refrigerated transport equipment are rather rigid especially as regards the following factors:

- efficiency and durability of insulation;
- adequacy and reliability of the cooling mechanism; and
- adequate circulation of air within the vehicle or container so that variations of temperature are slight.

Decisions on the establishment of new refrigerated transport services for eggs should be based on thorough economic as well as technical evaluations.

The following criteria should be taken into consideration.

- The need for a managerial and operational staff that is competent in all the operations involved in assembling, loading and distributing.
- The necessity of a sufficient volume of trade throughout the year.
- The possibility of making up loads with other compatible produce, e.g. dairy products.
- The possibility of carrying return loads, once eggs have been distributed.
- The degree to which the demand for refrigerated transport is concentrated geographically.

Marketing Organization for Eggs

The greater the distance between producer and consumer, the more complex is the marketing organization required to ensure that eggs reach consumers in the form, place and time desired. Producers may decide to market their produce directly to consumers-direct marketing-or may choose from a variety of marketing organizations that make up a marketing channel. Direct marketing includes the following methods of selling:

- sales from the farm (farm gate);
- door-to-door sales;
- producers' markets; and
- sales to local retail shops.

A typical marketing channel is made up of:

- collectors;
- assembly merchants;
- wholesalers; and retailers.

Direct Marketing

Egg producers who are situated a short distance from consumers may be able to practise direct marketing. Before choosing to sell their products directly to consumers, however, they must evaluate two main factors:

- *Time:* Producers who choose direct marketing have less time for production activities.
- *Cost:* The costs involved in direct marketing.

There are four main ways to carry out direct marketing.

Sales from the Farm

Producers may be able to sell eggs directly from the farm (farm gate). This, however, will depend on whether consumers are able and willing to go to the producer's facilities.

The main advantage of farm-gate selling is that the producer may be able to obtain a market price for eggs without incurring marketing costs. The main advantage for the consumers is that eggs will be fresh with little or no quality loss.

Door-to-door Sales/Street Hawking

Some consumers prefer that eggs be brought directly to their door. This means that the producer must spend time on marketing; however, consumers may appreciate the service and be willing to pay a good price. Furthermore, the producer can take orders directly from consumers and carry only what he/she is assured will be bought.

Producers' Markets

Usually the producer simply occupies a stall in a public marketplace and offers his/her produce for sale. Eggs are commonly displayed in baskets and often differentiated by weight/size and colour. Sales in producers' markets permit a farmer to make direct contact with consumers who are not able to go to the production facilities. The main disadvantage of using such markets is that, towards the end of the day, the producer may have to either reduce his prices sharply to dispose of remaining stock or carry it back to the farm.

Sales to Local Retail Shops

Producers can also sell directly to local retail shops. This requires some sort of agreement between the two parties regarding constant supply, quality and payment methods.

In some cases it may be possible for producers to sell directly to institutional consumers such as hotels, restaurants, schools and hospitals. This type of direct marketing, however, requires negotiation, which may result in a written contract of the duties and obligations of both parties. It also requires continual interaction over time between producer and buyer, a standard egg quality agreement and a constant supply. The producer must carefully evaluate the issues involved including the regular production and transport of large quantities of eggs.

Marketing Channels

A marketing channel is composed of a set of separate but interdependent organizations involved in the process of making a product available to consumers. The use of a marketing channel is convenient particularly when the producer does not have the time or financial means to carry out direct marketing. Intermediaries are usually able to make the product widely available and accessible because they are specialized and have experience and contacts. They also have a better understanding of the egg market. Intermediaries take the risks involved in marketing and also pay for the produce immediately.

Marketing Intermediaries

Collectors

Collectors undertake the initial work of assembling eggs from various producers or local country markets. They operate either on a commission basis or by purchasing on their own account. Where the quantity of eggs collected at each stop is small and frequent, this system is often the most economic. Collectors may be itinerant merchants, producers themselves, assembly merchants, wholesalers or their agents, or retailers.

Assembly Merchants

Assembly merchants may be divided into the following categories: local assembly market; independent processor-packer; and, cooperative processor-packer.

Local Assembly Market: In a typical local assembly market, a private firm, a producers' cooperative or a municipality provides an enclosed space for the use of sellers. Sales may take place by public auction or by private negotiation, subject to rules such as those on quality and payment arrangements. Auctioning requires the eggs to be graded and possibly presented in standardized containers, marked with identifying names or symbols. The local assembly market may provide cold storage facilities for the convenience of market users.

***Independent Processor-packer*:** This type of enterprise usually purchases eggs either through collectors or directly from producers.

The processor-packer may pass by the farm and pick up the eggs or the producer may deliver the eggs to the processing facilities where they are graded and packed. Usually eggs are sold to wholesalers; however, they are also sold directly to retailers and institutional consumers such as hotels, restaurants and hospitals.

Cooperative Processor-packer: The same type of enterprise may be set up and run by a cooperative association of producers. The main advantage is that the business is run by and for those who use it, rather than by those who own it. Cooperatives can obtain financing, provide extra competition to independent processor-packers and provide an alternative to established intermediaries.

Before forming a cooperative, producers should carefully evaluate:

- the market for eggs;
- problems in existing marketing channels and how to remedy them;
- the degree of know-how that producers have in marketing;
- rules and regulations;
- legal status;
- availability of finances;
- staffing requirements; and
- appropriate geographic location.

Wholesale Distributor

Wholesaling includes all the activities involved in selling goods to those who buy for resale or for business use. The main function of the wholesale distributor is to balance supplies against retail requirements and to take the initiative of bringing produce from areas where it is plentiful and cheap to those where it is relatively scare and expensive. Wholesalers usually have a good knowledge of the market, access to the best information on trends and prospects and working capital to carry business risks as required.

Wholesalers usually obtain eggs from central wholesale markets, assembly merchants, collectors and local country markets; however, in some instances they go directly to the producers. Eggs may be purchased directly or accepted for sale on a commission basis. Many wholesalers have their own storage facilities. Wholesale distributors may engage specialized transport agencies to transport eggs or operate such services on their own account.

Central Wholesale Markets

Central wholesale markets receive shipments from large farms and from country markets, and constitute a supply source where wholesalers and retailers can obtain the various types of produce they need. General wholesale markets sell many different products, including eggs.

Because it is the focus point of many smaller markets and also the point of contact for suppliers to important groups of consumers, a central market is usually the primary price-making mechanism for the production areas it serves. In this way it balances demand and supply.

Retailer

In urban areas, egg sales are made through retailers. Four types of retailers usually carry eggs in their shops:

- poultry shops where only eggs and poultry are sold;
- food shops specializing in eggs, poultry, cheese, butter, meat and fish;
- general food shops and supermarkets selling all kinds of foods and household goods; and
- meat markets where all types of meat are sold and eggs are also offered for sale.

In some instances retailers buy eggs directly from the producer and may have their own process-packing facilities. As we have seen, marketing channels have different organizations that carry out different functions, or it may be possible that an organization carries out more than one function. Vertical integration occurs when more than one of the stages of the marketing channel is carried out by a single organization. For example, a wholesaler may have processing-packing facilities, retail outlets and employ collectors as well.

Evaluation of The Marketing Channel

Before choosing a marketing channel or channels to market eggs, producers should carefully evaluate the following factors:

- market requirements and their ability to meet these;
- the type of intermediaries available;
- the number of intermediaries necessary to reach the market;
- alternative intermediaries different from the established marketing channel;

- the responsibilities of intermediaries and terms of possible agreements;
- costs involved;
- possible sales by the marketing channel; and
- the possibility of selling through a number of marketing channels.

Pricing and Sales Policy

Demand and Supply

The level of demand for eggs is determined by the price, the number of potential consumers, their purchasing power and by the extent to which they prefer to buy eggs rather than alternative foodstuffs. From the point of view of supply, a price must be high enough to cover production, storage and transport costs. It is unlikely that suppliers will continue to supply eggs if the price remains below that required to cover their costs and give them at least as high a standard of living as they could obtain in other ways. Hence in the long run, market prices must be both low enough for consumers to purchase and high enough to ensure that producers will supply.

Pricing

Usually market demand and supply determine egg prices. It is important for producers to ascertain market prices for eggs and the price trends over a one-year period. Once market prices are known, producers will be able to calculate if that price or prices in a market or various markets will cover their costs and give them a sufficient profit. It must be remembered that prices change and that pricing information must be up to date when calculating possible profits. In some countries there are seasonal variations in both production and demand which affect the level of prices at various times of the year. These variations in prices should be noted by the producer in planning his/her production and marketing.

Marketing Costs

Marketing costs will vary according to the method of marketing chosen. The main operating expenses for marketing include:

- packaging and storage;
- handling;
- transport;

- product losses;
- fees, taxes and unofficial payments, and
- unexpected costs.

Packaging and Storage Costs: Costs for packaging include the materials used for packaging, which may vary from a simple basket to a carton made of plastic, and labelling. The cost of storing the eggs must also be considered.

Handling Costs: The cost of packaging the eggs, putting them into storage, loading them for transport and unloading them at their destination must all be calculated as handling costs. Each individual handling cost may not amount to much; however, the sum total of all such handling costs can be significant.

Transport Costs: Costs for transport will vary according to the method of transport used and the distance covered.

Product Losses: Produce can be lost during the marketing period. There are two types of losses-quality and quantity. Eggs exposed to heat with consequent deterioration is an example of quality loss. Breakage of eggs during transport on a bumpy road is an example of quantity loss.

Fees, Taxes and Unofficial Payments: It may be that set fees have to be paid, for example, to a local authority for the use of a market stall. Taxes will have to be paid and, in some situations, bribes may be required to pass a roadblock or to access determined markets. These are all costs that must be considered.

Unexpected Costs: It is always important to calculate expenses for unexpected events that may raise costs. For example, it could happen that a road is closed and this may result in a longer distance to be covered to consign eggs. This will raise costs.

Marketing Costs

Costs	***US$***
Packaging/storage	
Handling	
Transport	
Product losses	
Fees, taxes, unofficial payment	
Unexpected costs	***Total costs***

At the end of the year, the producer can work out the production and marketing costs and the average market price for eggs over the year. After verifying how many eggs were sold during the year, the producer can calculate whether or not a profit was made.

Total Costs

Production costs	US$
Rearing	
Houses	
Equipment	
Feed	
Labour	
Vaccinations	
Mortality	
Various expenses	
	Total production costs
Marketing costs	
Packaging/storage	
Handling	
Transport	
Product losses	
Fees, taxes, unofficial payments	
Unexpected costs	
	Total marketing costs
	Total costs

Price Differences between Markets

One way of checking pricing efficiency in a marketing system is to compare the prices of similar qualities and types of eggs in different markets. Where the differentials reflect the necessary cost of some essential marketing service such as transportation, the marketing system can be regarded as fairly efficient. In other cases, it may be found that these differentials are larger than might be expected. Poor reporting and communication of market news as well as bad transport and storage facilities are among the most common causes of such discrepancies.

It is normal for prices to be lower in production areas than in deficit centres of consumption. Sometimes, however, the differential is much larger than the transport and other marketing charges would warrant. Lack of market information may make it difficult for wholesalers to judge how much produce a market will absorb and to estimate accurately the quantities that are being brought in by other buyers. Lack of storage facilities may be another contributory factor. Storage facilities would enable the wholesalers to move eggs in and out of storage to correct imbalances of supply and demand. Frequently, transportation between production and consumption areas is expensive, difficult to organize and risks heavy losses. Provided there is competition between traders in these markets, such price differentials should contract as these defects are corrected.

Seasonal Variations and Cyclical Movements

Seasonal changes in the prices of eggs mainly reflect variations in production. In temperate climates, the natural laying season is during the spring. Prices tend to be lower in the spring because of the plentiful supply and tend to be much higher in autumn when eggs are scarcer.

In climates where seasonal changes are less marked, variations in the availability of feed often cause fluctuations in marketing supplies of eggs. Producers should plan to get the eggs to market when prices are high. An important consideration in adjusting to such cycles is not whether prices are high or low by any particular standard, but whether other producers decide to expand or cut down their breeding flocks in response to them. The main aim is to secure a more even supply of eggs over the year at relatively stable prices.

Development of Sales Outlets

Most producers and traders are interested in expanding their markets. The simplest approach is to dispatch a selected lot of eggs to some consumer centre where prices appear attractive and find out by experiment whether the net return is greater than that obtained locally. If this proves successful, other consignments of eggs may be sent. Before large consignments are prepared for distant markets, it is recommended that market potentials be investigated. By undertaking some marketing research, the risk of losses can be minimized and the chances of developing profitable trade relationships greatly increased.

Investigation of Potential Markets

In planning a sales development programme the following points merit careful attention:

- available supply;
- potential markets;
- controls; and
- type and quality of product.

Available Supply: The number, type and quality of eggs possible to produce must be estimated realistically. A proportion of the eggs produced may not meet the quality standards desired. Seasonal and year-to-year variations in the supply are an important consideration.

Potential Markets: Potential markets should be investigated by looking at the following criteria:

- eggs sales in markets where there is a deficit for eggs;
- price levels throughout the year;
- total sales level achieved;
- distribution and its costs;
- competition; and
- consumer likes and dislikes.

Controls: Each potential market may have control restrictions such as minimum quality standards as well as packaging and disease controls. These must be investigated accurately. Also, there may be informal restrictions on new producers who want to enter a market and these have to be verified.

Type and Quality of Produce**:** In a potential market the type and quality of eggs required for that market must be assessed carefully. Marketing research can be used to determine the quality and form in which eggs are desired by consumers, and in what units and packaging they wish to buy them.

Estimates of how much consumption would change and in what direction if a shift in income or price occurred could also be verified.

Selling Arrangements: When a potential market has been located the next step is to establish trade contracts. It is important to select wisely the agent, distributor or retailer through which sales will be made.

One must ascertain the reliability, contacts and facilities of the person or organization involved. Importantly, a contract should be stipulated that clearly defines the duties and obligations of all parties concerned and the duration of the agreement.

Marketing Services

Market information, marketing education and training, promotional campaigns to promote egg consumption, marketing research to aid in producer and trader decision making and the availability of credit are all needed to help a marketing system operate more efficiently.

These activities may be seen as facilitating services for producers and traders. Marketing services include the following:

- Extension and training
- Market information services
- Marketing research
- Programmes to expand consumption
- Trade associations
- Credit.

Extension and Training

Those involved in production and marketing of eggs should engage regularly in training. The broad objectives of most egg marketing educational programmes are to help producers understand the demands of the market and modify their production and marketing accordingly. Processors/packers, wholesalers and retailers can be helped to become more effective and efficient so that eggs can be marketed with less waste, less loss in quality and at a lower cost.

Extension officers can lead meetings, discussions and demonstration programmes on egg production and marketing. They should make regular visits to production and marketing centres to keep in touch with current developments and problems. The extension officer can provide a valuable link between technical research workers and market intermediaries and the producers.

The duties of an extension officer are described below:

1. *Understand the functioning of the egg industry.* This involves looking at such issues as egg production processes, statistics, major enterprises, the geographic distribution of egg production, price trends, sales volumes, sales methods and when sales occur.
2. *Advise farmers on the possible potential of egg production.* Advice can be given on what marketing opportunities there may be, how to calculate the demand for eggs, how to calculate marketing

and production costs and what processing and storage facilities may be required.

3. *Raw materials.* The extension worker must advise the farmers on where to obtain equipment and materials for building brooder and laying houses, feed, small chicks and all other materials that are necessary for production. Materials needed for packaging must also be considered. Importantly, the farmer must be advised on how to grow and manage small chicks, the feed required and the type of environment necessary. The input suppliers should be surveyed and their prices for equipment, small chicks, feed, etc., collected. Delivery and credit conditions with various suppliers should be covered. The extension worker should advise small farmers to group together to purchase and transport raw materials to the various small farms. This will result in cost savings for the farmers.
4. *Financing of raw materials.* Extension workers should determine how farmers could finance the required inputs. Can farmers rely on their own cash savings or credit institutions? If not, what credit is available and from which institutions, and is it possible to promote savings that may then be invested in egg production.
5. *Production.* The farmer needs to be advised on what breed to buy, when to buy day-old chicks, when to place grown chicks into the laying house, when they will start laying and how long they will lay profitably. The production cycle should be covered thoroughly and all requirements, such as feed, water, clean nests, etc., should be included.
6. *Post-production facilities.* Farmers should be advised on facilities that are required once eggs are produced, such as storage facilities, and cleaning, grading and packing facilities. Information concerning the cost of such facilities and where materials can be obtained should be provided.
7. *Promote small farmer associations.* Practical advice on the formation of cooperative or group production, packaging, processing and sales associations and pooling schemes should be given. Importantly, the extension worker should promote to farmers the idea of grouping or associating together. This will lead not only to savings in the purchase of inputs, but will also improve opportunities for egg marketing. The pool of raw materials, production and marketing capabilities will create

a better bargaining position for small-scale farmers. It will also enable them to have better access to credit, and will give them the opportunity to adopt innovations more easily and at a lower cost.

8. *Understanding marketing.* Farmers must be assigned to understand what marketing is and what are the marketing channels for eggs. Different prices may be obtained from different markets. Farmers must be active in looking for buyers and in determining who they are, what price they may obtain and quantities of supply required and, furthermore, whether they pay in cash and when they pay, and whether the price they pay is higher or lower than that of other buyers. Farmers must learn how to calculate their production and marketing costs. They must also be able to understand when and where to sell eggs and the quantities to be sold based on market information. It is also important to understand the costs and possible profitability of storage.
9. *Pricing.* Farmers should be advised on the principal factors that form and influence prices. They should learn how to calculate costs and profit.
10. *Marketing channels.* Extension workers should constantly monitor the channels available. They should explain the channels, their efficiency and costs to farmers and advise on possible channel alternatives. Furthermore, they should teach farmers how to monitor channels and explain the opportunities that may arise from using different channels and the relative cost savings that may be obtained.
11. *Legislation.* The extension workers should explain to the farmers the legislation that could affect production and marketing of eggs, relative quality standards, sales contracts, etc.
12. *Sources of market information.* Farmers should be told how to obtain market information from government, local municipalities, radio bulletins, etc. Farmers should be trained to carry out simple marketing research.
13. *Challenges and opportunities.* Possible risks and opportunities that may be present in the industry in months or years to come must be considered by the extension worker. They should advise on the need for smaller packaging, new production and processing techniques that may allow for cost savings, new market openings, etc.

14. *Requirements to improve marketing.* The extension worker should instruct farmers on a regular basis regarding the prevention of losses during handling and transport, standards, quality control, grading methods and candling, simple but effective cooling devices, etc.
15. *Visits.* Regular visits to packaging, grading and processing establishments should be made by the extension worker so that he can constantly monitor the situation. Where possible, extension workers should encourage farmers to visit production and processing facilities.

Market Information Services

The importance of market information has to be emphasized. This information is of vital importance for producers and traders. It will enable them to produce and trade based on what markets require. Market information could be defined as a service, usually operated by the public sector, that involves the collection on a regular basis of information on prices, and in some cases quantity, of widely traded agricultural products from rural assembly markets, wholesale and retail markets.

This service also involves the dissemination of this information on a timely and regular basis, through various media, such as radio and newspapers, to producers, traders and consumers. Up-to-date reports on supplies available, quantities sold and in storage, prices paid at major markets at local, wholesale and retail levels are invaluable for an efficient marketing system. Marketing information services can help in the following ways.

- *Improve bargaining between producers and traders.*
- *Risk reduction.* Producers who have reliable and timely information, and who can interpret it, for example, can decide to which market they want to send their eggs in order to maximize returns. Information reduces transaction costs by reducing risks.
- *Identification of markets.* It is unlikely that producers and traders will consign eggs to a distant market unless they are reasonably confident of being able to sell at a profit. Market information can help in taking such a decision.
- *Allocation of productive resources.* Information on market price fluctuation over a period of time can help a producer decide

whether to expand, contract or keep production constant. This information allows the producer to allocate production resources more efficiently.

- *Storage decisions.* Egg storage implies costs; therefore, producers and traders need to obtain a price that covers possible storage costs. Information on seasonal price trends is important for producers and traders.
- *Trade development.* Marketing information not only alerts producers to production possibilities, but can also give information on trading opportunities. This can lead to an increase in market outlets for the producers and make them more competitive.
- *Facilitating contractual agreements.* Contractual agreements between producers and traders usually carry a set price agreement for a period of time for the eggs supplied. Market information can help set a fair price in the contractual agreement and thus avoid disputes.

Assembling reliable, valid and unbiased reports is not an easy task. Experience in interpreting and checking information supplied by individual buyers and sellers is essential. The collection of market information is especially difficult when many transactions take place through private negotiations, yet it is here that it is most needed. Personal enquiries of buyer and seller may be necessary. The transactions covered should be those that have the most influence on price making and which concern the most important categories of eggs traded. Care should be taken to relate prices to quality, implying the use of a uniform set of specifications throughout the reporting sequence.

Marketing Research

Marketing research is necessary in order to help producers make decisions regarding marketing. Marketing research can be defined as the systematic and objective search for, and analyses of, information relevant to the identification and solution of any problem in the field of marketing. Marketing research aids decision-making, however, it will not fully eliminate risk.

Marketing research has advantages and disadvantages. The main advantages of marketing research are:

- defining the needs and nature of customers and their ability and desire to buy;

- scanning the business environment;
- gathering needed information for decision-making;
- reducing risk;
- helping in production planning; and
- monitoring and controlling marketing activities.

Marketing research should be carefully planned and each step of the process analysed before the actual research begins. The first step is to clearly define the purpose of the research and the objectives. The objectives must be measurable, quantifiable and attainable. Costs of the research must be carefully evaluated and budgeted and the time duration considered.

Possible marketing research plan for eggs:

1. *Recognition of information needed or problems.* It is important to define carefully the information that is required. For example, from a simple information requirement such as what eggs to sell, many sub-questions may arise such as the following.
   - What type of eggs is most in demand?
   - Are more brown eggs sold?
   - Are more white eggs sold?
   - What size egg is sold the most?
   - Do different outlets require eggs of different sizes, colour and packaging?
   - What is the market price?
   - Who is buying eggs?
2. *Definition of objective(s).* The objective or objectives of the research should be defined clearly. For example, do shopkeepers want eggs in trays or in retail cartons?
3. *Deciding on what tool(s) to use to gather information.* Information may be gathered by observation, survey, or from already published and available data. Observation is simply observing phenomena and recording them as they occur, for example, observing consumers at a market and what they buy. A survey involves questioning consumers, wholesalers and retailers in order to gain information of interest. A survey also involves preparing a questionnaire. For example, consumers may be asked what type of eggs they prefer, while retailers could be asked what type of eggs they sell the most.

4. *Formulation of appropriate tool(s).* Importantly, if either observation or survey is chosen as a tool to collect information, the tool must be carefully designed. The design for observation should tell the observer if he or she must look at a particular aspect of a market or observe the whole market. The design for the survey should look at what type of questions should be asked and how they should be asked. For example, the questions could be as follows.
   - Do you like eggs? Yes/No
   - Do you buy eggs? Yes/No
   - Do you buy eggs every day? Yes/No
   - Do you buy eggs at least once a week? Yes/No
5. *Information gathering.* When the information is being gathered, it must be done in an unbiased and uniform manner and be recorded accurately.
6. *Analysis of information.* Once the information has been gathered, it must be analysed and evaluated. For example, if consumer data has been gathered regarding egg buying habits, it is necessary to group the various results into categories, such as income and/or geographic location of customers.
7. *Results and conclusions.* The summarized data will give some clear results from which conclusions may be drawn. For example, if it was found that the majority of consumers eat brown medium-sized eggs, one should aim to produce brown medium-sized eggs.

Programmes to Expand Consumption

The main methods that can be used to expand consumption are improved marketing organization and consumer education and promotion.

Improved Marketing Organization

Improvements in marketing organization and methods are often a prerequisite to expanding consumption and hence production.

The existence of well-run buying, packaging/processing and distribution enterprises promotes both production and consumption. It dispels doubts in the minds of potential consumers regarding quality, freshness and wholesomeness of the eggs offered.

Consumer Education and Promotion

A continuous programme of consumer education and promotion may help increase egg consumption. Eggs are of high nutritional value, easily digestible and especially good for children, pregnant women and the elderly. Educating young school children regarding the goodness of eggs may prove over time to be a good strategy to increase consumption. Radio advertising, radio talk shows, advertising, collaboration with public and private agencies seeking to improve health, living and nutritional standards may all increase consumption. Such programmes can be carried out by single enterprises or enterprises grouped together in order to share the costs and benefits of the initiative. Importantly, such initiatives should be carried out in collaboration with retail outlets, schools, hotels, restaurants and hospitals so as to obtain the maximum benefit of the educational and promotional campaign.

Trade Associations

It may be of great value to form a trade association for egg producers. An association can impose determined standards and requirements in order to guarantee that eggs produced by its members are of a determined quality. The association may help in introducing and disseminating knowledge and new techniques of production and marketing. It may put producers into good bargaining positions *vis-à-vis* transport facilities, wholesalers and retailers.

Trade associations are set up on a voluntary basis by enterprises and are usually most effective when organized by individuals and enterprises with common business interests. Membership is voluntary and funds are obtained either as a fixed amount per year from each member or as a fixed amount on volume sold. Usually officers are elected to perform association duties and permanent or *ad hoc* committees may be appointed to handle certain issues or programmes.

Credit

Producers' credit needs have an important bearing on marketing organization and costs. Limited access to credit is a common barrier to the establishment of improvements and increased production and marketing. For example, if the establishment of new production facilities such as hen houses is required, the producer may need to obtain credit. In order to acquire eggs and finance their movement through marketing channels, traders must either draw the necessary capital from their own resources or be able to obtain it on short-term loan.

Marketing Success in Chile

The Chilean egg industry is predominately based on the traditional family business. The industry has about 8 million layers in production on an annual basis. Two million layers are found in large production facilities, while the greater part, 6 million layers, comes from small-scale producers.

In Chile, between 75 and 80 percent of all eggs marketed are white shell eggs, the remaining are brown. Ninety-two percent of egg sales are through public markets and small retail stores, while the remaining 8 percent are sold in supermarkets.

About 80 percent of egg producers are associated with Asohuevo, Chile's producer association. This association carries out strong publicity campaigns and managed to increase per capita consumption from 105 eggs in the 1980s to 165 in the 1990s.

Bibliography

Andrew L. Winton: *Poultry Eggs*, Agrobios, Delhi, 2002.

Appleby, M.C & Hughes, B.O. & Elson, H.A.: *Poultry Production Systems, Behaviour, Management and Welfare,* CAB International, NY, 1992.

Berton, V. and Mudd, D.: *Profitable Poultry: Raising Birds on Pasture,* USDA's Sustainable Agriculture Network (SAN), Washington, DC, 2001.

Blackbourn, David: *The Long Nineteenth Century: A History of Germany,* Oxford University Press, New York, 1998.

Butterworth, A.: *Measuring and Auditing Broiler Welfare,* CABI, UK, 2004.

Chenoweth, H.: *Free-Range Poultry,* Free-Range Poultry Production and Marketing, Creola, Ohio, 2001.

Dekkers, J.C.M., H.H. Zhao, and RL. Fernando: *Linkage Disequilibrium Mapping in Livestock,* Belo Horizonte, Brazil, 2006.

Dekkers, J.C.M.: *Methods and Strategies for QTL Mapping,* University of Wisconsin, Madison, 2005.

Elson, H.A.: *Poultry Production Systems, Behaviour, Management and Welfare,* CAB International, NY, 1992.

Foreman, P.: *The Chicken Tractor: The Permaculture Guide to Happy Hens and Healthy Soil-All New Straw Bale Edition,* Good Earth Publications, USA, 2002.

Grist, A.: *Poultry Inspection, Anatomy, Physiology and Disease Conditions,* Nottingham University Press, UK, 2006.

Hatta, H.: *Hen Eggs: Basic and Applied Science,* CRC Press, Delhi, 1996.

Hill, Donna : *Biosecurity in the Poultry Industry,* International Book, 2006.

Homer O. Stuart: *Commercial Poultry Farming,* Biotech Books, Delhi, 2011.

Huopalahti, R. & Lopez-Fandino, R. & Anton, M. & Schade, R.: *Bioactive Egg Compounds,* Springer, Delhi, 2007.

Jordan, F & Pattison, M & Alexander, D & Faragher, T.: *Poultry Diseases,* WB Saunders, UK, 2002.

Jull, Morley A. : *Successful Poultry Management*, Biotech, Delhi, 2001.

Keith Wilson N.D.P: *A Handbook of Poultry Practice*, Agrobios, Delhi, 2000.

Leclercq, B.: *Nutrition and Feeding of Poultry*, Nottingham University Press, U.K., 1994.

Mandal, A.B. : *Nutrition and Disease Management of Poultry*, International Book Distributing Co, Delhi, 2004.

McNab, J.M. & Boorman, K.N.: *Poultry Feedstuffs, Supply Composition and Nutritive Value,* CABI Publishing, UK, 2002.

Owen, W Powell : *Poultry Farming and Keeping*, Biotech Books, Delhi, 2005.

Pathak, M R : *Nutrition and Disease Management of Poultry*, International Book Distributing Co, Delhi, 2004.

Powell., C. : *Microbiological and Hydraulic Evaluation of Immersion Chilling for Poultry*, Journal of Food Protection, 1995.

Prasad, L N : *Advanced Pathology and Treatment of Diseases of Poultry : With Special Reference to Etiology Signs*, International Book Dist, Delhi, 2006.

Rama Rao, S V and M R Reddy: *Growth Promoters in Poultry : Novel Concepts*, International Book Dist, Delhi, 2008.

Randall, C.J.: *Color Atlas of Diseases and Disorders of the Domestic Fowl and Turkey,* Iowa State University Press, UK, 1991.

Scanes, C.G. & Brant, G. & Ensminger, M.E.: *Poultry Science,* Pearson Prentice Hill, New Jersey, 1992.

Stadelman, W.J. & Cotterill, O.J.: *Egg Science and Technology,* Food Products Press, Imprint of Haworth Press, New York, London, 1995.

Tullett, S.G.: *Poultry Science Symposium Number 22.* Butterworth-Heinemann, NJ, 1991.

Verma, S. S.: *Microbiological Changes on Chicken Carcasses During Processing*, Indian Journal of Poultry Science, 1989.

Weeks, C & Butterworth, A.: *Measuring and Auditing Broiler Welfare,* CABI, UK, 2004.

Whitehead, C.C.: *Bone Biology and Skeletal Disorders in Poultry,* Carfax Publishing Company, U.K., 1992.

Wilson, G.C.: *Egg Quality Handbook,* NSW DPI, MT, 1990.

Wiseman, J. & Garnsworthy, P.C.: *Recent Developments in Poultry Nutrition,* University Press, India, 1999.

Yamamoto, T. & Juneja, L.R. & Hatta, H.: *Hen Eggs: Basic and Applied Science,* CRC Press, Delhi, 1996.

Young, M.: *Controlling Newcastle Disease in Village Chickens,* ACIAR, US, 2002.

Index

A

B

C

E

❑❑❑